Key Geography Places

David Waugh
Former Head of Geography
Trinity School
Carlisle

Tony Bushell
Former Head of Geography
West Gate Community College
Newcastle upon Tyne

With Chris Peyton and Alasdair McLelland

Stanley Thornes (Publishers) Ltd

Text © David Waugh, Tony Bushell, with Chris Peyton and Alasdair McLelland, 1995, 2000
Original line illustrations © Stanley Thornes (Publishers) Ltd 1995, 2000

First edition designed and typeset by Hilary Norman
Second edition typeset by Clare Park
Scottish edition typeset by Hardlines
Illustrations by Hardlines, Lovell Johns Limited/Maps International, Angela Lumley, Tim Smith and Shaun Williams
Photo research by Julia Hanson and Penni Bickle

The previous page shows favelas and skyscrapers in Rio de Janeiro

First published in 1995 and reprinted seven times.
Second edition published in 2000
Scottish edition published in 2000 by:
Stanley Thornes (Publishers) Ltd
Delta Place
27 Bath Road
Cheltenham GL53 7TH
United Kingdom

00 01 02 03 04 / 10 9 8 7 6 5 4 3 2 1

A catalogue record for this book is available from the British Library.

First edition ISBN 0-7487-1942-3
Second edition ISBN 0-7487-5439-3
Scottish edition ISBN 0-7487-4188-7

Printed and bound in China by Dah Hua Printing Press Co. Ltd.

Acknowledgements

The authors and publishers are grateful to the following for permission to reproduce photographs and other copyright material in this book.

AA Photo Library (pp. 67 (5)); Air Images (p. 126); Allsport (p. 38 B); Andes Press Agency (pp. 6 bottom right, 15 centre bottom, 32 centre left, 61 C); The Automobile Association (p. 78 A); Benetton UK (p. 74 logo); Cephas (pp. 67 (2), 70 B, 88 B, 93 G, 97 C, 98 centre and bottom left); John Cleare Mountain Photography (pp. 41 D, 67 (3)); Bruce Coleman (pp. 45 bottom, 50 B); Sue Cunningham Photography (pp. title page, 15 bottom, 28 C, 32 bottom left); Eye Ubiquitous (pp. 94 B, 99 bottom, 101 B); Fiat UK Limited (p. 74 logo); Frank Lane Picture Agency (p. 44 top); Frank Spooner (pp. 15 centre top, 76 B, 77 D); GeoScience Features (p. 64 top); Highlands and Islands Enterprise (p. 131 F); Honda (UK) (p. 108 C); Hutchison Library (pp. 6 top right, centre left, centre right, 26 A, 32 centre right, 67 (5)); Image Bank (pp. 5 left (both), 6 top left, centre left, bottom left, 67 (1), 76 top left, 77 top left); Images of Africa (pp. 38 C, 39 D, 50 A, 52 C, 54 A, 58 C); Images Colour Library (pp. 5 right, 93 D, 96 A, 98 top); Intermediate Technology (pp. 6 centre right, 13 D, 49 C, 54 B, 57 B and C); Japan Information and Cultural Centre (p. 98 bottom right); Japan National Tourist Organisation (pp. 93 E, 99 top); Johnstons of Elgin (p. 131 top left); Newslink Africa (p. 5 centre); Olivetti UK Limited (p. 74 logo); Maps reproduced from Ordnance Survey mapping with the permission of the Controller of Her Majesty's Stationery Office © Crown Copyright (pp. 121, 127, 133); Panos Pictures (pp. 4 bottom left, 25 top, 32 bottom right, 49 top); Chris Peyton (pp. 124 A, 125 D); Pirelli SpA of Milan (p. 74 logo); Port of Tokyo, Tokyo Metropolitan Government (p. 95 D); Press Association (p. 80 top); Rex Features (pp. 4 top left, 12 C, 53 D, 80 bottom, 92 C); Robert Harding Picture Library (pp. 14 bottom, 90 C); Science Photo Library (pp. 16, 38 A, 94 A, 108 B); Scotland in Focus (pp. 112 B, C) /A G Johnston (p. 117 C) /Barnes (p. 119 D) /Davies (p. 125 top) /Firth (p. 131 E) /Moar (p. 122 top right) /Robbeson (p. 112 bottom right) /Schofield (p. 120) /Weir (p. 118 A) /Whitehorne (pp. 112 top left, 124 B); Scottish Parliament, Holyrood Project Team (p. 122 bottom left); John Seely (p. 57 D); Skishoot-Offshoot (p. 72 A); South American Pictures (pp. 15 top, 20 C and D, 21 E, 24 top, 27 D, 28 A, 29 E); Spectrum Colour Library (pp. 52 A, 68 A, 72 B, 92 B, 93 F, 94 C, 96 B, 101 C); Sporting Pictures (p. 14 centre); Still Pictures (pp. 14 top left, 24 right, 25 bottom left and right, 28 B and D, 32 top left and right, 39 E, 44 centre, 50 C); Sygma (pp. 26 B, 64 bottom, 89 D and E, 90 D); John Taylor (p. 122 top left, bottom right); Tony Stone Worldwide (pp. 20 A, 21 G); David Waugh (pp. 40 A and B, 41 E, 49 D, 52 B, 53 E, 54 C); Judith Waugh (p. 48 top); The Western Isles Tourist Board (p. 124 logo); World Pictures (pp. 44 bottom right, 45 top, 68 B); Zanussi Ltd (p. 74 logo).

Every effort has been made to contact copyright holders. The publishers apologise to anyone whose rights have been inadvertently overlooked, and will be happy to rectify any errors or omissions.

Contents

1 Development

Differences in world development

All countries are different. Some, for example, are rich and have high standards of living. Others are poor and have lower standards of living. Countries that differ in this way are said to be at different stages of development.

Map **A** shows the distribution of these countries. Notice that the richer countries are mainly in the 'North' and the poorer countries in the 'South'.

Activities

1 Complete a larger copy of table **B**.

B

	Continent	Developed or developing?
Brazil		
Italy		
Japan		
Kenya		
UK		

2 Complete a copy of the crossword by solving the following clues. All the answers can be found on these two pages.

Across
1 Country with a population density of 19 per km².
2 London is its capital city.
3 Continent between the Atlantic and Indian Oceans.
4 Ocean between South America and Africa.
5 Tropic to the south of the Equator.
6 Country in North America.
7 Developed country in southern Europe.
8 The capital city of Italy.
9 Developed country in North America.

Down
5 Tropic to the north of the Equator.
10 Developing country in Asia.
11 African country on the Equator.
12 Country with a population density of 322 per km².
13 Ocean between Asia and North America.
14 Developed country south of the Equator.
15 Capital city of Kenya.
16 Capital city of Brazil.

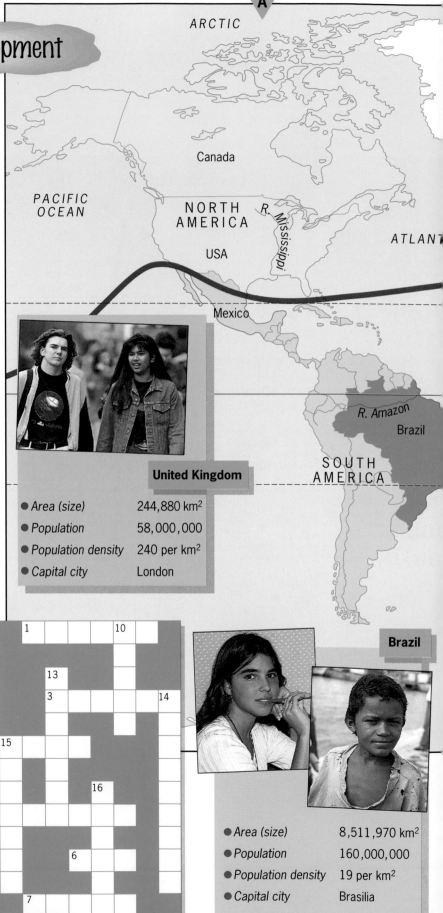

A

ARCTIC

Canada

PACIFIC OCEAN

NORTH AMERICA

R. Mississippi

USA

ATLANT

Mexico

R. Amazon

Brazil

SOUTH AMERICA

United Kingdom

- Area (size)   244,880 km²
- Population   58,000,000
- Population density   240 per km²
- Capital city   London

Brazil

- Area (size)   8,511,970 km²
- Population   160,000,000
- Population density   19 per km²
- Capital city   Brasilia

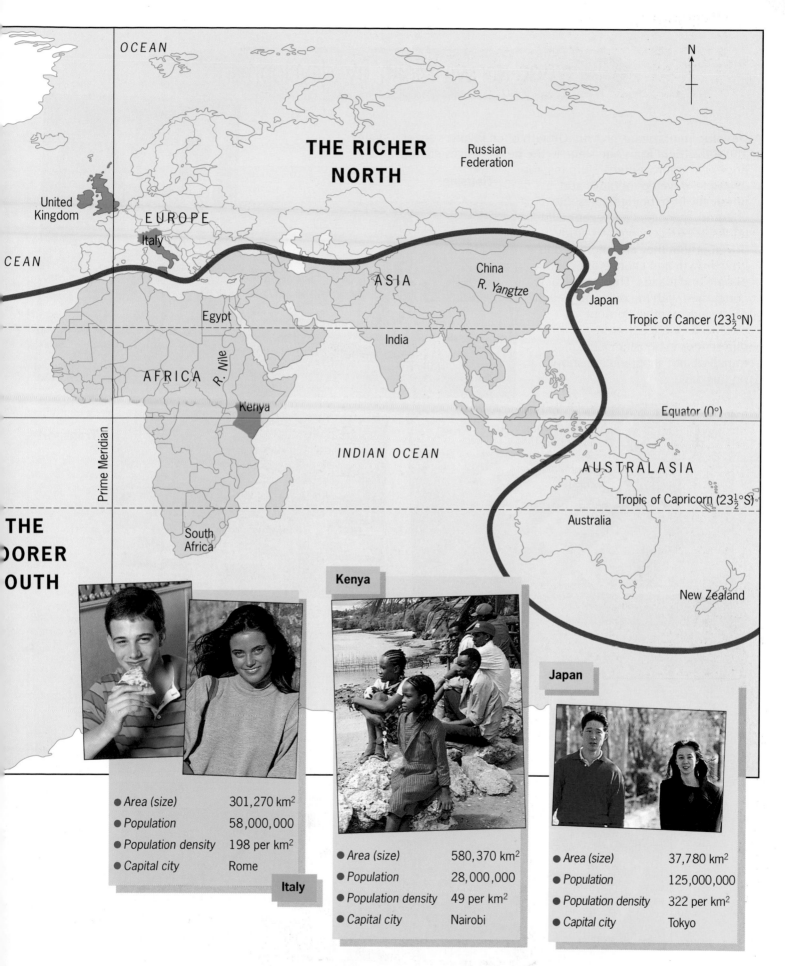

THE RICHER NORTH

Russian Federation

OCEAN

N

United Kingdom

EUROPE

Italy

OCEAN

ASIA

China
R. Yangtze

Japan

Egypt

Tropic of Cancer (23½°N)

India

AFRICA

R. Nile

Kenya

Equator (0°)

Prime Meridian

INDIAN OCEAN

AUSTRALASIA

Tropic of Capricorn (23½°S)

Australia

THE
POORER
SOUTH

South Africa

New Zealand

Kenya

Japan

- Area (size) 301,270 km²
- Population 58,000,000
- Population density 198 per km²
- Capital city Rome

Italy

- Area (size) 580,370 km²
- Population 28,000,000
- Population density 49 per km²
- Capital city Nairobi

- Area (size) 37,780 km²
- Population 125,000,000
- Population density 322 per km²
- Capital city Tokyo

5

What do we mean by development?

Development means growth. Growth often involves several stages. The movement from one stage to the next involves change and, hopefully, improvement. Humans, for example, develop mentally and physically from being babies into children, teenagers and eventually adults. When geographers study development they are interested in how growth and change affect both people and places. They are also concerned with the differences that each stage of development produces. Some extreme differences in housing, jobs, transport and shopping are shown on this page.

Housing

Jobs

Transport

Shopping

It is not easy to measure development. The usual and the easiest method is by measuring wealth (diagram **A**). The wealth of a country is given by its **gross national product per capita** (**GNP**). GNP is the total amount of money earned by a country divided by the total number of people living within that country (diagram **B**). To make comparisons between countries easier, GNP is given in US dollars (US $). Diagram **C** gives the GNP per capita for several countries.

Based upon wealth, people try to fit countries into one of two groups.
1 The richest countries are said to be **economically more developed** (**developed**). Their inhabitants have high standards of living.
2 The poorest countries are said to be **economically less developed** (**developing**). Their inhabitants have much lower standards of living.

There are problems in using GNP as a measure of wealth. The main one is that wealth is not shared out evenly within a country. This problem is greatest in developing countries where there is a huge gap between the rich and poor. In other words, the rich are very rich and the poor are very poor.

A To most people, development means how rich or how poor a country is. People who live in rich countries usually have a high standard of living.

B In 1995 the UK earned $1 094 000 million. If you divide this by the UK's total population of 58 million this gives a GNP per capita of $18 700.

C Countries in the richer 'North' have high GNPs, e.g.
Japan $39 640
USA $26 980
Italy $19 020
UK $18 700

Countries in the poorer 'South' have much lower GNPs, e.g.
Brazil $3 640
Kenya $280
India $340

Activities

1 Match these beginnings to the correct endings.

2 Work in pairs. Use the photos on page 6 and your own knowledge to make a list of differences between **developed** and **developing** countries.

3 Map **D** divides the world into the 'rich' developed countries and the 'poorer' developing countries. Into which group do the following countries fit?
Brazil • India • Italy • Japan • Kenya • UK • USA

Development is	a low standard of living
A developing country has	the growth of places
A developed country has	the measurement of a country's wealth
Standard of living is	a high standard of living
GNP per capita is	people's level of wealth

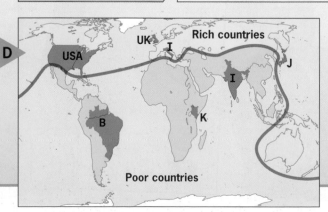

D Rich countries
UK
USA
I
J
I
B
K
Poor countries

Summary

There are considerable differences in development between countries. One way of measuring these differences is through wealth.

How can population be used to measure development?

Wealth is not the only way to measure, or describe, development. Development can be described by using **social** measures such as **population**. Countries at different stages, or levels, of development have different **birth rates**, **death rates**, **natural population growth**, **infant mortality rates** and **life expectancy**. These are explained in diagram **A**.

Look at table **B**. It uses the five development measures described in diagram **A**. It also gives the actual figures for these measures for five countries. These countries are the four that are referred to in this book plus the United Kingdom. The UK figures have been added to make the comparisons more meaningful. Notice the differences between the countries, especially those between Japan or Italy and Brazil or Kenya. The differences are mainly due to the various countries being at different stages of development.

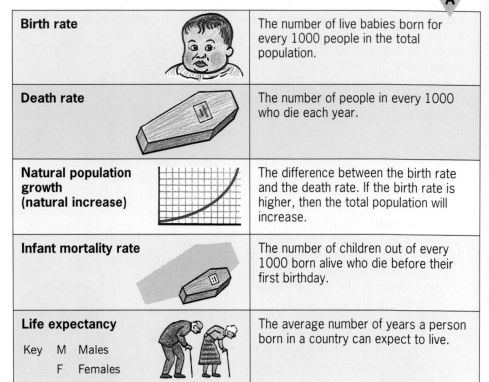

A

Birth rate		The number of live babies born for every 1000 people in the total population.
Death rate		The number of people in every 1000 who die each year.
Natural population growth (natural increase)		The difference between the birth rate and the death rate. If the birth rate is higher, then the total population will increase.
Infant mortality rate		The number of children out of every 1000 born alive who die before their first birthday.
Life expectancy Key M Males F Females		The average number of years a person born in a country can expect to live.

B

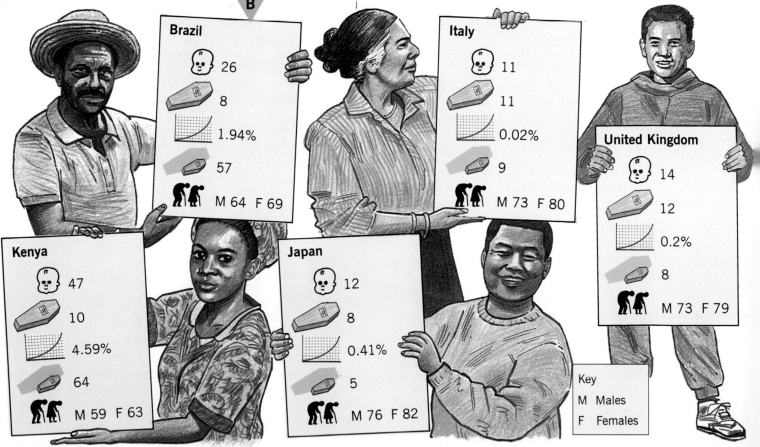

Brazil
- 26
- 8
- 1.94%
- 57
- M 64 F 69

Italy
- 11
- 11
- 0.02%
- 9
- M 73 F 80

United Kingdom
- 14
- 12
- 0.2%
- 8
- M 73 F 79

Kenya
- 47
- 10
- 4.59%
- 64
- M 59 F 63

Japan
- 12
- 8
- 0.41%
- 5
- M 76 F 82

Key
M Males
F Females

Graphs **C** and **D** are **population pyramids**. They show:
- the population of a country divided into 5-year age groups;
- the percentage (proportion) of males and females in each age group.

They are useful because they
- allow comparisons to be made between countries;
- show the present population structure of a country (and so measures its development);
- help to forecast future population structures (in order to find out what problems and needs there might be in the future).

Graph **C** has a shape which is typical of many developing countries. Graph **D** is more typical of a developed country.

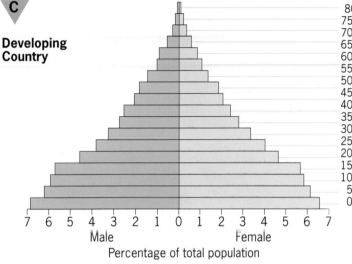

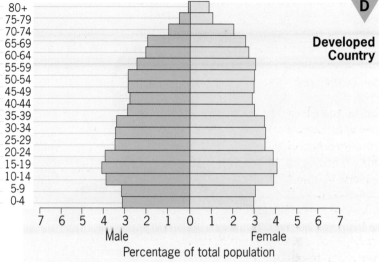

- A high birth rate – a large number of children aged under 15
- A high infant mortality rate
- A narrowing pyramic suggesting a high death rate
- Many people in the 'reproductive age group' (20–39) suggesting more children and a rapid natural increase in population
- A shorter life expectancy means fewer people reach old age

- A low birth rate – relatively few children aged under 15
- A low infant mortality rate
- A straighter pyramid suggesting a lower death rate
- Fewer people in the 'reproductive age group' (20–39) suggesting fewer children and a slow natural increase in population
- A longer life expectancy means more people reach old age

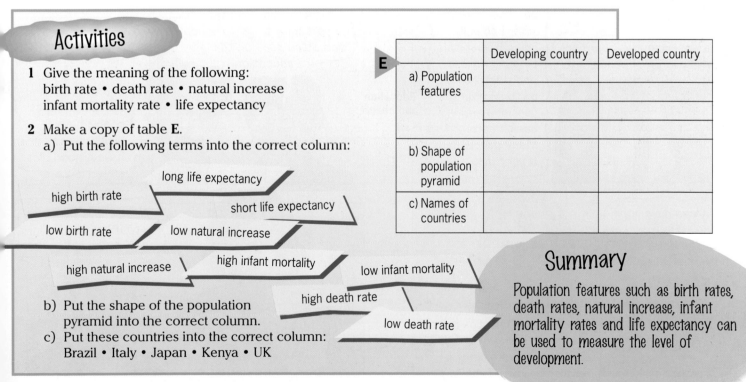

Activities

1 Give the meaning of the following:
birth rate • death rate • natural increase
infant mortality rate • life expectancy

2 Make a copy of table **E**.
   a) Put the following terms into the correct column:

long life expectancy
high birth rate
short life expectancy
low birth rate
low natural increase
high natural increase
high infant mortality
low infant mortality
high death rate
low death rate

   b) Put the shape of the population pyramid into the correct column.
   c) Put these countries into the correct column:
   Brazil • Italy • Japan • Kenya • UK

		Developing country	Developed country
E	a) Population features		
	b) Shape of population pyramid		
	c) Names of countries		

Summary

Population features such as birth rates, death rates, natural increase, infant mortality rates and life expectancy can be used to measure the level of development.

How else may development be measured?

Apart from wealth (GNP) and population there are many other ways of trying to measure the level of a country's development. Most of these are **social** measures. Some of these alternatives are described in diagram **A** and are listed in table **B**. Table **B** also gives actual figures for these measures for the four countries described in this book, together with those for the UK.

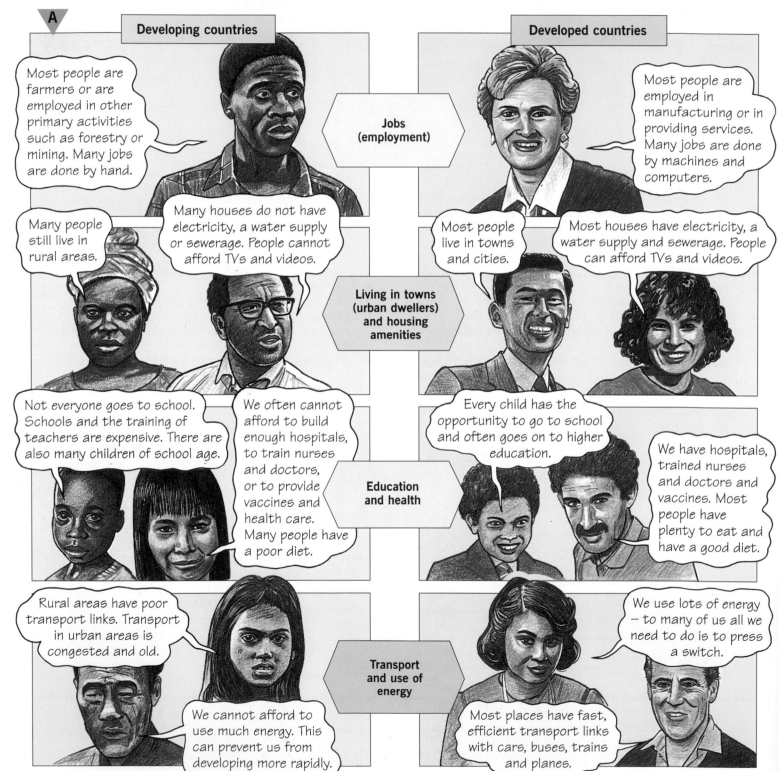

B

	Country				
	Brazil	Italy	Japan	Kenya	UK
Working in agriculture and primary jobs (%)	29	9	7	81	2
Working in manufacturing (%)	16	20	24	7	20
Urban dwellers (live in towns) (%)	83	72	78	32	94
Literacy (adults who can read and write) (%)	81	97	99	69	99
People for each doctor	1600	210	610	10130	300
Calories per person (food consumption)	2730	3498	2921	2064	3270
Cars per 1000 people	70	472	280	5	358
TV sets per 1000 people	213	424	620	9	435
Energy used per person (kg coal equivalent per 1000 people)	44	265	450	10	487

C

The Human Development Index (HDI)

The United Nations have replaced GNP, an economic measure of development, with the Human Development Index (HDI). The HDI is a social-welfare index. It measures literacy, life expectancy and infant mortality The HDI League Table for 173 countries has

-
  1 Japan;
  15 UK;
  22 Italy;
  70 Brazil;
  129 Kenya.
-

The HDI can also be used to show differences in levels of development within a country as well as between countries.

Activities

1 Table **D** lists 15 measures which can be used to show development.
   a) Make a larger copy of the table.
   b) For each measure, rank the countries in order. The first measure, for GNP, has been done for you.
   c) Total up the scores for each of the five countries. According to these measures the country with the **lowest** total is the most developed. The country with the **highest** score is the least developed.
   d) Which country appears as the most developed? Write a paragraph to describe some of the features of a developed country.
   e) Which country appears to be the least developed? Write a paragraph to describe at least five features of a developing country.

D

	Gross National Product (US$)	Lowest birth rate	Lowest death rate	Lowest natural increase	Lowest infant mortality	Longest life expectancy	Highest % in urban areas	Lowest % in agriculture and primary jobs	Highest % in manufacturing	Highest literacy rate	Fewest people per doctor	Most calories per person	Most cars per 1000 people	Most TVs per 1000 people	Highest energy use	TOTAL
	(See page 7)															
Brazil	4															
Italy	2															
Japan	1															
Kenya	5															
UK	3															

Summary A wide range of measures can be used to show differences in the level of development between various places in the world.

The need for interdependence and sustainable development

Interdependence

Most countries want to be **independent**. They would prefer to make their own decisions as to how they develop and how their people live. However, no country can live in isolation. No country has everything that it needs. This means that each country has to work with other countries if it is to develop and improve its standard of living. Countries which work with, or rely upon, other countries are said to be **interdependent**.

One of the ways by which countries become interdependent is through **trade**. Trade is when countries sell and buy goods from each other. Goods may either be **primary** or **manufactured**. Primary goods are **raw materials** such as minerals and foodstuffs. They are often low in value. Manufactured goods, such as

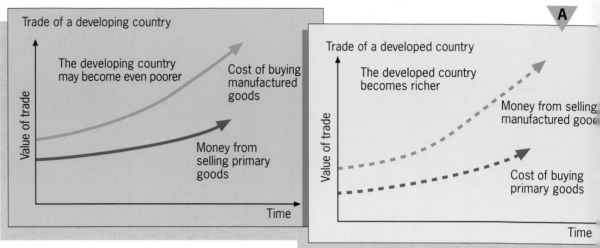

Trade of a developing country

The developing country may become even poorer

Cost of buying manufactured goods

Money from selling primary goods

Value of trade

Time

A

Trade of a developed country

The developed country becomes richer

Money from selling manufactured goods

Cost of buying primary goods

Value of trade

Time

machinery and vehicles, have a higher value. Developing countries usually have to rely upon selling primary goods, while developed countries rely upon selling manufactured goods (diagram **A**).

Unfortunately, as diagram **A** shows, trade can result in the richer, developed country becoming even richer and the poorer, developing country becoming even poorer. In time the developing country may have

to ask the developed country for **aid** (diagram **B**). If so, it is likely to have to rely more and more upon the developed country. In this way it will become even more interdependent and less independent.

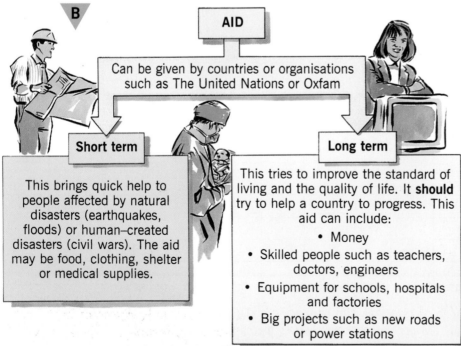

B

AID

Can be given by countries or organisations such as The United Nations or Oxfam

Short term

This brings quick help to people affected by natural disasters (earthquakes, floods) or human–created disasters (civil wars). The aid may be food, clothing, shelter or medical supplies.

Long term

This tries to improve the standard of living and the quality of life. It **should** try to help a country to progress. This aid can include:

• Money
• Skilled people such as teachers, doctors, engineers
• Equipment for schools, hospitals and factories
• Big projects such as new roads or power stations

C Food aid in Somalia

Sustainable development

It is very easy for countries that want to develop and to make progress, to waste resources and damage the environment. The richer countries are the worst offenders. With only 25 per cent of the world's population, they use 75 per cent of the earth's natural resources. Many of these resources are non-renewable and can only be used once. They often cause pollution of air, land and water. The poorer countries add to the problem. They have rapidly rising populations who want to improve their standards of living. They rarely have enough money or the technology to provide or use additional resources without damaging the environment.

Many people believe the solution to the problem is **sustainable development**. Sustainable development (diagram D) is progress that can continue year after year. It uses, but does not waste, natural resources. It improves, but does not threaten, ways of life. It neither harms the present, nor destroys the future environment. It is sensible development which can bring benefits now and in the future.

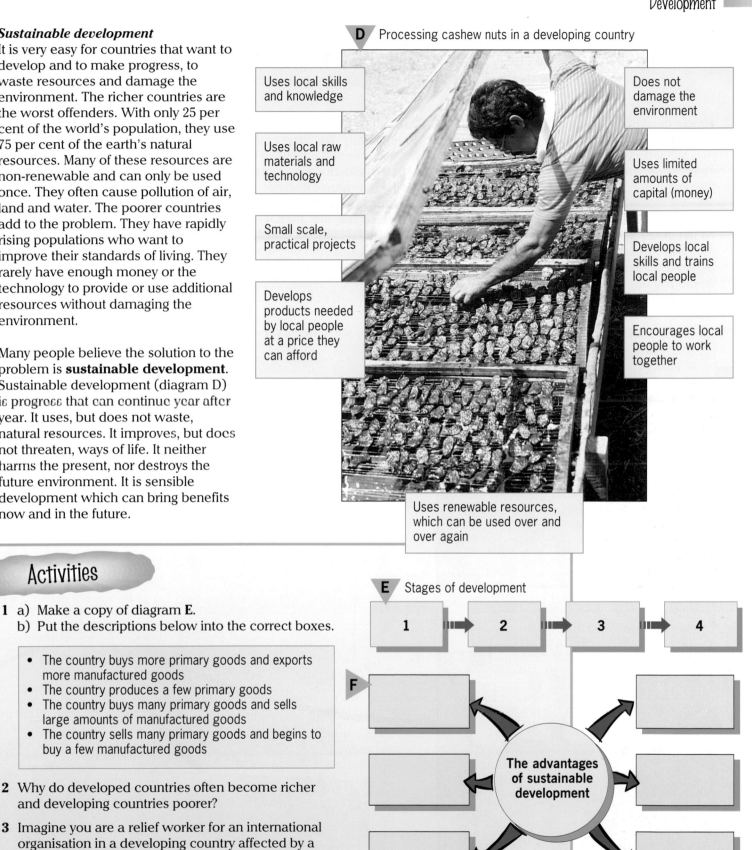

D Processing cashew nuts in a developing country

Uses local skills and knowledge

Uses local raw materials and technology

Small scale, practical projects

Develops products needed by local people at a price they can afford

Does not damage the environment

Uses limited amounts of capital (money)

Develops local skills and trains local people

Encourages local people to work together

Uses renewable resources, which can be used over and over again

Activities

1 a) Make a copy of diagram **E**.
   b) Put the descriptions below into the correct boxes.

- The country buys more primary goods and exports more manufactured goods
- The country produces a few primary goods
- The country buys many primary goods and sells large amounts of manufactured goods
- The country sells many primary goods and begins to buy a few manufactured goods

2 Why do developed countries often become richer and developing countries poorer?

3 Imagine you are a relief worker for an international organisation in a developing country affected by a severe earthquake. Write a letter to your organisation's head quarters in the UK. Your letter should describe the type of aid which the country needs most urgently.

4 Make a copy of diagram **F**. Complete it by adding six advantages of sustainable development.

E Stages of development

1 → 2 → 3 → 4

F

The advantages of sustainable development

Summary Countries need to be interdependent if they are to develop and share in the earth's resources.

What is Brazil like?

Almost everybody has heard of Brazil but few people have much idea of what Brazil is really like. What do you know about Brazil? What images come to mind when you think about the country?

Most people, when asked that question, would probably come up with football, coffee, the River Amazon, Rio de Janeiro and perhaps Indians and the rainforest. Which of these would you have thought about? Would you have added any others?

Brazil in fact, is very difficult to describe. It is a huge country. It is bigger than all of Europe put together and almost 30 times larger than Great Britain. Because of its size it has a great variety of physical features, climate and types of vegetation. There are also great variations in how people live, their wealth and overall quality of life. Indeed Brazil can best be described as a land of contrasts. Look at the photos on these pages. They show some of the contrasts.

Amerindians live a traditional way of life in the hot, wet rainforest.

A mixture of people from all over the world now live in Brazil. The team that won the 1998 football World Cup is typical of that mix.

Brazil is one of the fastest growing industrial nations in the world. Industry has brought wealth but it has also caused pollution.

Living conditions can be very difficult. These stilt houses are on the banks of the Amazon.

Peasant farmers try to make a living in the hot, dry semi-desert of the north-east.

Poverty and unhealthy living conditions are common in Brazil. These slum houses are in Salvador.

For rich and poor alike the beach is the place to be for social life, sport and leisure. This is Copacabana beach.

So, Brazil is huge, exciting, interesting and colourful. It has forests and semi-deserts. It has spectacular cities and areas of unexplored wilderness. It has Indian tribes and a population made up of people from all over the world.

But it is the contrast between the rich and the poor that people notice more than anything else. On the one hand there is the appalling poverty of the slums. On the other there is the high-class life-style of the wealthy. That is Brazil's great problem and the one that is causing the most concern.

Activities

1 Work with a partner and make a list of things that come to mind when you think of Brazil.

2 From the photos and their captions find:
   a) one place that is hot and wet and one place that is hot and dry
   b) one place that is jungle and one place that is semi-desert
   c) two examples of wealth and two examples of poverty
   d) one example of traditional lifestyles and one example of mixed nationalities
   e) one example of difficult working conditions and one example of economic wealth.

Write your answers in a table like the one below.

Contrasts in Brazil		
a) Climate		
b) Vegetation		
c) Standard of living		
d) Population		
e) Economic		

3 a) Describe at least four things that you think are good about Brazil. Write a paragraph of about 50 words.
   b) Make a list of at least six things that you think are problems in Brazil.

Summary Brazil is a land of contrast. There are great differences between how the rich and the poor live.

What are Brazil's main physical features?

River Amazon

- Second longest river in the world
- 6280 km (3925 miles) in length
- Over 300 km wide at its mouth
- More than 1100 tributaries flow into it
- Has source in Peru at height of 5200 m (17000 ft)
- Ocean going ships can sail 3200 km up river
- Contains 20% ($\frac{1}{5}$ th) of world's total river water

Amazon rainforest

- Largest rainforest on earth
- Mainly flat and low-lying
- A few hills and low mountain ranges
- Over 40000 species of plants and animals
- Home to over a third of all species on earth
- Mostly uninhabited
- Home of Amerindians
- Still large areas unexplored
- Always hot and wet
- Rains on most days

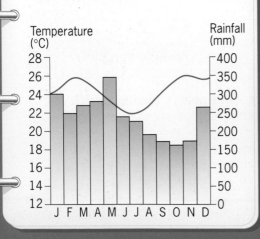

Coastal lowlands

- Highlands fall steeply into the sea
- Some narrow, flat stretches of land
- Area once rainforest
- Forest now cleared for farming and settlement
- Warm temperatures, ample rainfall, good soils – excellent for growing crops
- Inland areas cooler than coast
- East coast is the most populated area of Brazil

The photos that make up this picture of South America were taken from 830 km (520 miles) above the earth's surface by an orbiting satellite. The colours have been changed slightly (**enhanced**) to show the relief features and drainage patterns more clearly. The red dashed line has been added to show Brazil's border.

Dense forest and areas of lush vegetation show up as bright green. Drier regions are brown or yellow. Notice the river Amazon and its tributaries. Can you also see the snow-capped mountains of the Andes in the west of South America?

North-east Brazil

- Mostly flat-topped highland
- Some areas of flat coastal lowland
- Very dry. No rain at all in some years
- Mainly rough scrub and thorn bushes
- Too dry for much else to grow
- Vegetation called caatinga
- Difficult living conditions

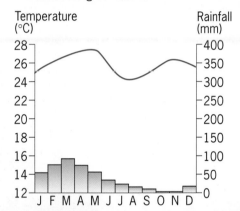

Brazilian Highlands

- Mainly flat-topped hills and plateaux
- Height between 1000 and 2000 metres
- Include highest parts of Brazil
- Hot, wet summers
- Warm, dry winters
- Grassland and open forest

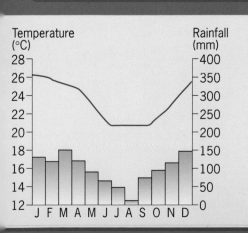

Activities

1 Make a larger copy of the map below. Complete the descriptions of each region using information from these two pages.

Brazil – main physical regions

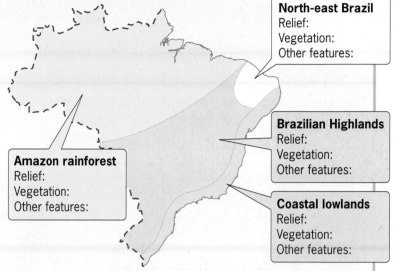

North-east Brazil
Relief:
Vegetation:
Other features:

Brazilian Highlands
Relief:
Vegetation:
Other features:

Coastal lowlands
Relief:
Vegetation:
Other features:

Amazon rainforest
Relief:
Vegetation:
Other features:

2 Match the following regions with the correct climate descriptions. Use the climate graphs to complete the temperature and rainfall figures for November.

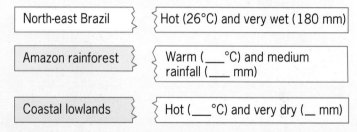

| North-east Brazil | Hot (26°C) and very wet (180 mm) |

| Amazon rainforest | Warm (___°C) and medium rainfall (___ mm) |

| Coastal lowlands | Hot (___°C) and very dry (__ mm) |

3 Write out the following paragraph and fill in the missing spaces.

The River Amazon
The Amazon is the _____ longest river in the world (the Nile is the longest river). It flows ___ km from its source in ____ to its mouth in the _____ . Ocean-going ships can sail ___ km upriver. At its _____ the Amazon is over 300 km wide. The river has over ____ tributaries. It drains 6.5 million km² of rainforest and is important for transport and trade.

Summary

Brazil is a huge country with a great variety of physical features. Much of the land is flat and there are few mountains. The Amazon is Brazil's greatest river.

People in Brazil

Look at diagram **A** which shows some people who live in Brazil. Notice how different each of them look and what varied backgrounds they have. If you stand on a street corner in a Brazilian city you will see even more variety. Nearly every race, culture and religion in the world will be represented.

Just 500 years ago Brazil was populated only by tribal Indians. Today it is a mixture of people from all over the world. This change has come about because of the large number of migrants who have arrived in Brazil and the mixing that has gone on between different races. The

Portuguese first came to what we now know as Brazil in 1500. They ruled the country as a colony for over 300 years. They brought slaves from Africa to work on plantations and helped develop tolerant attitudes towards new citizens from foreign countries. The slaves were quickly followed by immigrants from other parts of Europe, Asia and North America.

Most modern Brazilians are proud of their mixed origins. There is much inter-marriage and this is one reason why there are very few racial problems and practically no colour prejudice in Brazil.

A

We are the original Amerindians. Most of us live in the Amazon forest but there aren't many of us left now

We come from Japan. Most of us live in and around São Paulo. We like to keep our traditions and tend to live together and keep to ourselves.

We Portuguese were the first Europeans to settle here. We set up modern Brazil and gave the country our language.

We Germans have been coming here since the 1880s. We have brought industry and business skills to Brazil.

Our people first came here as slaves over 400 years ago. Our background and African culture are an important part of Brazilian life.

We are Italians. Some of our people went straight to the cattle ranches in the south but most took work on the coffee plantations near São Paulo.

Brazil has a population of 159.1 million – almost three times that of the UK. So where do they live and why are some parts of Brazil very crowded whilst others are almost uninhabited?

Map **C** shows **population distribution** in Brazil. You will notice that most people live near the coast in the south-east of the country. Towards the north and west the **population density** is very **sparse**. The reasons for this uneven distribution are due to **positive** and **negative factors**.

B

Negative factors discourage people from living in an area
• Difficult living conditions
• Poor farming
• Few resources available

Positive factors attract people to an area
• Pleasant living conditions
• Good farming
• Resources available for industry

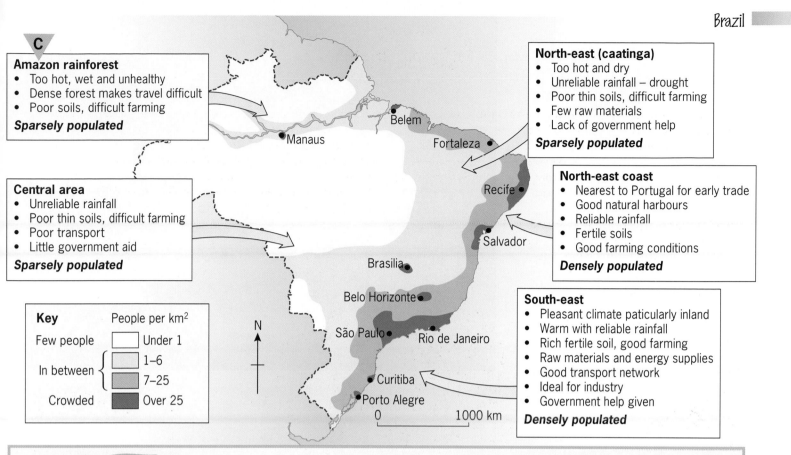

C

Amazon rainforest
- Too hot, wet and unhealthy
- Dense forest makes travel difficult
- Poor soils, difficult farming

Sparsely populated

North-east (caatinga)
- Too hot and dry
- Unreliable rainfall – drought
- Poor thin soils, difficult farming
- Few raw materials
- Lack of government help

Sparsely populated

Central area
- Unreliable rainfall
- Poor thin soils, difficult farming
- Poor transport
- Little government aid

Sparsely populated

North-east coast
- Nearest to Portugal for early trade
- Good natural harbours
- Reliable rainfall
- Fertile soils
- Good farming conditions

Densely populated

South-east
- Pleasant climate paticularly inland
- Warm with reliable rainfall
- Rich fertile soil, good farming
- Raw materials and energy supplies
- Good transport network
- Ideal for industry
- Government help given

Densely populated

Key	People per km²
Few people	Under 1
In between	1–6
	7–25
Crowded	Over 25

N

0   1000 km

Cities on map: Belem, Manaus, Fortaleza, Recife, Salvador, Brasilia, Belo Horizonte, São Paulo, Rio de Janeiro, Curitiba, Porto Alegre

Activities

1 a) Who were the original inhabitants of Brazil?
 b) Name five countries or regions from which immigrants came to Brazil.
 c) Which people gave their language to Brazil?
 d) Which people arrived as slaves?
 e) Which people prefer to keep together?

2 Which three of the following are reasons for there being little racial discrimination or colour prejudice in Brazil?
- All the people are of the same race
- The Portuguese encouraged tolerance
- Everyone is the same colour
- There is much inter-marriage
- Brazilians are proud of their mixed origin.

3 Draw table **D** and put the following in the correct columns:

• densely populated • sparsely populated • very few people • crowded • too hot and wet • pleasant climate • good water supply • raw materials nearby • dense forest• good energy supplies • difficult travel• unhealthy conditions • much industry • rich fertile soil • poor soils • difficult farming • good farming • mainly negative factors • mainly positive factors.

D

South-east	Amazon rainforest

4 a) List the cities from sketch **E** in order of size. Give the biggest first.
 b) In which part of Brazil are the four largest cities?

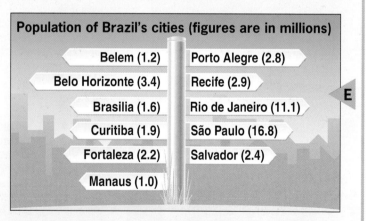

Population of Brazil's cities (figures are in millions)

Belem (1.2)	Porto Alegre (2.8)
Belo Horizonte (3.4)	Recife (2.9)
Brasilia (1.6)	Rio de Janeiro (11.1)
Curitiba (1.9)	São Paulo (16.8)
Fortaleza (2.2)	Salvador (2.4)
Manaus (1.0)	

E

5 Draw a bar graph to show the number of people in each of the cities in sketch **E**.
- Arrange the bars in order of size with the biggest on the left.
- Use different colours for the cities that are inland to those that are within 200 km of the coast.
- Give your graph a title.

Summary

Brazil is a mixture of people from all over the world. The south-east is the most densely populated. Population decreases towards the north and west.

19

The best of Brazil

For many people, Brazil is a dream holiday destination. A pleasant climate, a varied landscape, and friendly people are just a start. Add to this sun-clad beaches, historic sights and the adventure of the Amazon jungle, and Brazil has everything for the tourist. One way to see Brazil is by a multi-centre holiday. Here is one such holiday advertised by a tour company as 'The best of Brazil'.

A

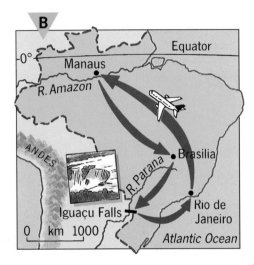

B

C

The Rio Carnival Four magical days of noise, excitement, music and non-stop dancing. A festival of street parties, parades, fancy dress and firecrackers. Endless wine, food and fun with very, very friendly people. This is Brazil at its best, a never to be forgotten experience.

Day 1 Rio
Excursion to the summit of Corcovado mountain and the statue of Christ the Redeemer. Spectacular views of the bays and beaches. Afternoon free to visit Copacabana and Ipanema beaches or stroll along elegant avenues of shops and boutiques.

Day 2 Rio
Morning tour of the city including ascent of Sugar Loaf Mountain by cable car. Optional cruise to one of the nearby tropical islands. Evening at a top nightclub and dancing to the 'Samba beat'.

Day 3 Rio/Manaus
Early flight to Manaus. Travel downriver by canoe to experience the thrill of the Amazon (**E**). A chance to see thousands of colourful birds leave their roosts and fish jump from the waters of the river. An evening expedition by boat searching for alligators by torchlight (**D**).

Day 4 Amazon
In the morning explore the rainforest with a local guide. There are more plants and animal species here than anywhere else in the world. Over 42,000 different species of insects have been recorded so be well prepared with insect repellent! In the afternoon visit an Indian village to experience the local way of life and perhaps buy traditionally made crafts.

Day 5 Amazon
Canoe trip to Lake Acajituba to see rubber trees, Brazil nuts and variety of palms. In the afternoon a fishing trip. There are 1500 species of edible fish here. You may even catch piranha.

Day 6 Manuas/Brasilia
Morning flight to Brasilia. Afternoon city tour to view the exciting modern architecture of this ultra modern capital.

Day 7 Brasilia/Iguaçu
Fly via São Paulo to Iguaçu (**F**). Guided tour of one of the world's most spectacular waterfalls. Don't miss the thrilling helicopter flight which takes you right over the Falls.

Day 8 Iguaçu
A full day of relaxation at Iguaçu Falls set in a sub-tropical forest full of exotic birds and butterflies. Opportunity to cross the border to visit Paraguay and Argentina.

Day 9 Iguaçu/Rio
Morning flight to Rio. Day at leisure in city.

D

E

F

Postcard from
Das Cataratas Hotel, Iguaçu Falls

Dear Paul and Christine,

We're at Iguaçu Falls now. They are absolutely breathtaking. We have just walked along a narrow pathway carved out of the rocks and we're right beside the falls. There are over 275 separate waterfalls here. They empty a million gallons of water every second into the foaming Parana River. The sound is deafening. We are going on a boat trip later to Garganta do Diabo (Devil's Throat). That's below the falls so we might get a bit wet! Some of our friends have booked the helicopter trip.

With best wishes
Joanne

G

Activities

1 List four features that make Brazil an interesting holiday destination.

2 Draw a star diagram like the one below to show at least six attractions of Rio de Janeiro.

H

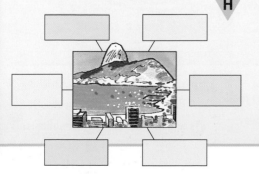

3 Imagine that you are on Day 5 of the Best of Brazil holiday. Write a postcard to a friend describing the most enjoyable parts of your holiday so far.

4 Imagine that you are flying from Iguaçu to Rio on Day 9. Write in your diary how you spent your time at Iguaçu and what impressed you about the falls.

5 Working in pairs, design a poster advertising Brazil as a holiday destination.
- Try to include a map, labelled photos or sketches, and short descriptions.
- Make the poster as colourful and interesting as possible.
- Your teacher may have some travel brochures which could help you.

Summary Brazil is a land of contrasts. It has varied and often spectacular scenery and a way of life that is both interesting and exciting.

Brazil's 'Golden Triangle'

The south-east is where most people in Brazil live. It is the richest area of the country and has always been the most important industrial region. Because of this, the three main cities of São Paulo, Rio de Janeiro and Belo Horizonte have become known as the '**Golden Triangle**'.

The south-east region is made up mainly of highland that slopes steeply into the sea. On the narrow, coastal plain there is little space for development and the climate is uncomfortably hot and wet. Inland on the high plateau the climate is much cooler and healthier. It is here where most industry and **commercial farming** is located.

Coffee provided early wealth for the region. The climate and rich 'terra rossa' soil proved perfect for growing this crop and São Paulo quickly developed as the 'coffee capital of the world'. Santos became important as the port for São Paulo. Immigrants flocked to the area to work on

the coffee plantations called **fazendas**. Over half a million were Italian. Others came from different parts of Europe and other parts of Brazil, mainly in the 1890s.

The profit from coffee was put to good use. Much of it was invested in other industries and on good roads and railways. At first the new industries were small in scale and linked to local resources. Factories processing food and producing leather goods, textiles, clothes and paper were the most common.

More recently, the mining of nearby mineral resources, the building of new power stations and an increase in government aid has helped the Golden Triangle develop **heavy industries**. Brazil now has steel, chemical, shipbuilding, aircraft and car industries all located in the area. These industrial developments have attracted even more migrants to the area in search of work.

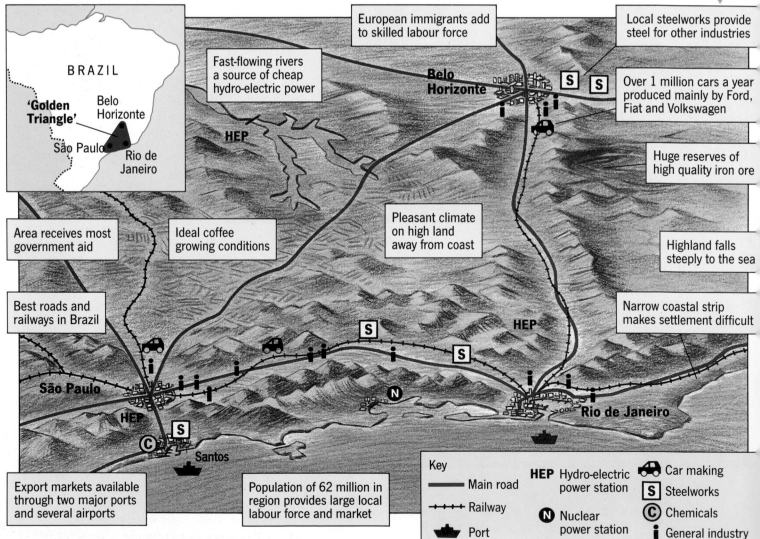

European immigrants add to skilled labour force

Local steelworks provide steel for other industries

Fast-flowing rivers a source of cheap hydro-electric power

Over 1 million cars a year produced mainly by Ford, Fiat and Volkswagen

Huge reserves of high quality iron ore

BRAZIL

'Golden Triangle'

Belo Horizonte

São Paulo

Rio de Janeiro

Pleasant climate on high land away from coast

Area receives most government aid

Ideal coffee growing conditions

Highland falls steeply to the sea

Best roads and railways in Brazil

Narrow coastal strip makes settlement difficult

São Paulo

Rio de Janeiro

Santos

Export markets available through two major ports and several airports

Population of 62 million in region provides large local labour force and market

Key

― Main road

+++ Railway

Port

HEP Hydro-electric power station

Ⓝ Nuclear power station

Car making

Ⓢ Steelworks

Ⓒ Chemicals

General industry

The success of the south-east has brought many benefits but it has also caused some problems . . . B

Only the rich people have made money from the factories. Most of the workers are paid very low wages.

Cars and factories pollute the air in our cities

Our rivers, lakes and even the sea are polluted by waste from towns and factories

Traffic congestion is a problem here. There are just too many cars.

Wildlife has been lost as forests have been cleared to make way for factories and more coffee plantations

Millions of people live in poor quality housing with no water, electricity or proper sanitation

As more people come to the area in search of work, there are not enough jobs to go round and unemployment increases

Many of the largest industries are multi-nationals owned by foreigners. This means that most of the profits leave Brazil.

Activities

1 Draw a simple sketch map of the region shown in map **A** as follows:
   a) Draw in the coastline and name the Atlantic Ocean.
   b) Mark and name São Paulo, Belo Horizonte, Rio de Janeiro and Santos.
   c) Draw in and name the 'Golden Triangle'.
   d) Shade the narrow coastal strip green and shade the highland brown.

2 a) Make a larger copy of diagram **C** below.
   b) Write the following labels in the correct places: Santos, narrow coastal plain, highland, mineral resources, much industry, hot and wet, pleasant climate, coffee growing.
   c) With help from your diagram suggest why fewer people live on the coast than inland.

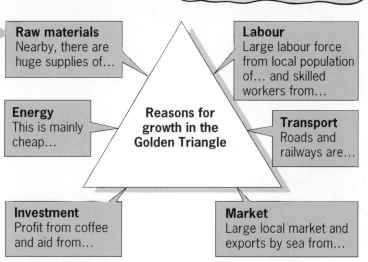

D **Raw materials** Nearby, there are huge supplies of…

Labour Large labour force from local population of… and skilled workers from…

Energy This is mainly cheap…

Reasons for growth in the Golden Triangle

Transport Roads and railways are…

Investment Profit from coffee and aid from…

Market Large local market and exports by sea from…

3 Make a larger copy of diagram **D**, above, and complete it using information from map **A**.

4 Study diagram **B** and list the problems of Brazil's south-east. Write no more than ten words for each one. Underline those that are damaging to the environment.

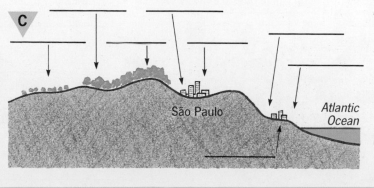

C

São Paulo   Atlantic Ocean

Summary

The south-east is Brazil's wealthiest and most populated region. The Golden Triangle has many industries and attracts large numbers of migrants in search of work.

São Paulo: the rich...

A

Hello,

my name is Maria and I live in São Paulo with my parents and brother. My father is a manager at the Ford car factory on the outskirts of town. My mother works in an office. We have a nice house in the suburbs with a large garden. A maid comes in on most days to cook the meals and keep the house clean.

My brother Carlos is two years older than me and we go to school together. I like geography and languages and hope to go to University eventually. During the week I work hard at school and I have a lot of homework.

At weekends we often go to our small beach house at Ubatura, a beach resort on the Atlantic coast. If we don't go there I like to go downtown.

São Paulo city centre is like any modern city. It has huge skyscrapers, lovely shops, lots of restaurants and plenty of nightlife. Rua Augusta is the best for fashion shops. My brother sometimes takes me to a disco near there. I like to dance the Samba.

São Paulo is the biggest and wealthiest city in Brazil. It is very crowded and always seems to be busy. The area around the city is the most important in Brazil for agriculture and industry. Over 90% of Brazilian-made cars are manufactured here. New factories and offices seem to open up nearly every day.

Like many families in São Paulo we are well off and have a good standard of living. Not everyone in our city is so lucky though. Some people here are very poor indeed and live a very difficult life.

Activities

1 Unscramble the following to give five features of São Paulo's city centre.

> prakysrecss • sphso • trustearna eflingthi • sdoisc

2 Why is the area around São Paulo important?

3 a) What jobs do Maria's mother and father have?
   b) Maria's family is wealthy. Give five features of her way of life that show this.

4 Make a larger copy of sketch **C** showing Roberto's house. Complete it by answering the questions.

C   Life in a favela

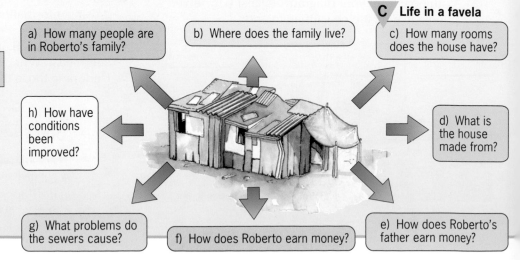

a) How many people are in Roberto's family?

b) Where does the family live?

c) How many rooms does the house have?

d) What is the house made from?

e) How does Roberto's father earn money?

f) How does Roberto earn money?

g) What problems do the sewers cause?

h) How have conditions been improved?

24

...and the poor

B

Hello,

I'm Roberto and am 12 years old. I live with my parents and four sisters on the outskirts of São Paulo. Our home is in a slum district known in Brazil as a favela. Our house is very basic. We built it ourselves with any materials that we could find. There are only two small rooms. In one we work, eat and relax. In the other we sleep. It is very crowded but we are making improvements all the time.

My father makes sandals from old car tyres. He sells them at a small market on the edge of our favela. On a good day he can make over one cruzado (about 80p). I only go to school in the mornings. There are very few books and there never seems to be any pens or paper to work with. In the afternoons and weekends I go to the city centre to work as a shoe-shiner. I will have to give up school soon and work full-time. My family needs the money.

The smell of our favela is very unpleasant. Open sewers run down the streets and even under houses. The place is dirty and disease is a problem. My two youngest sisters nearly died from diarrhoea when they were just babies. The council tries to improve conditions here. They recently provided us with piped water and electricity which is a great help. Nearly half of São Paulo's people live in slum conditions like us.

5 Copy and complete the paragraph on the right using the words from drawing **D**.

D

unhealthy   education   electricity
sewage   favela   poor

A slum district in Brazil is called a _____. It is an area of _____ quality housing which often lacks _____, a water supply and _____ disposal. People in these areas earn little money, have poor _____ and live in crowded, _____ conditions. *Favelas* are mainly the result of large numbers of newcomers moving to cities from the countryside.

Summary

Standards of living vary considerably in the São Paulo area. Some people are wealthy and have a good quality of life. Other people are extremely poor and live in slum conditions.

People in the Amazon rainforest

A

In the north and west of Brazil is the Amazon rainforest. This is the least populated region of the country and is the home of many different Amerindian tribes. The Kayapo are one of these tribes. They live on the Xingu river, a tributary of the Amazon.

Photo **A** shows a traditional Kayapo village. The circle of wooden buildings with palm-leafed thatched roofs are where the families live. The central house is used only by men. They meet socially here and make the decisions that affect the whole village.

Like most Amerindian people, the Kayapo use face and body paint. Red and black are the most common colours. Both of these are made from the juice of plants. On special occasions feathered head-dresses, nose pieces and ear plugs are worn. The men wear a round wooden disc between the chin and a stretched lower lip (photo **B**). All of the ornaments are made from materials found in the forest.

The Kayapo live by hunting, fishing and collecting food from the forest (photo **D**). A few crops are grown but only just enough for the villagers themselves. People who live like this are called **hunter-gatherers** and **subsistence farmers**. The forest provides the Indians with an excellent diet. They have a choice of over 250 fruits and hundreds of different vegetables, nuts and leaves.

B

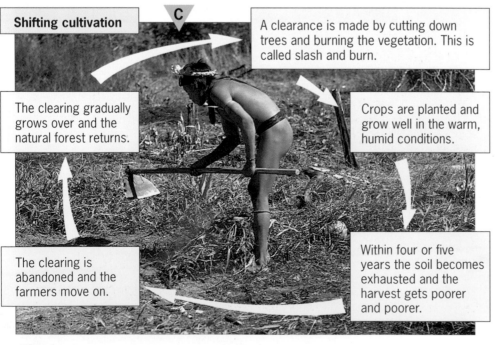

Shifting cultivation

C

A clearance is made by cutting down trees and burning the vegetation. This is called slash and burn.

Crops are planted and grow well in the warm, humid conditions.

Within four or five years the soil becomes exhausted and the harvest gets poorer and poorer.

The clearing is abandoned and the farmers move on.

The clearing gradually grows over and the natural forest returns.

As well as hunting and gathering food, the Kayapo make their own 'gardens' in small forest clearings. In these they grow fruit trees and crops including manioc, maize, sweet potatoes, pineapples, peppers and beans. The Indian women, especially, are expert forest gardeners. They work in a careful and organised way, improving the soil at the same time. After a while, however, the heavy rains wash away the nutrients in the soil and crops no longer grow so well. The Kayapo then move away and make a new **forest garden**. This type of farming is called **shifting cultivation** (photo **C**).

D

Forest uses

- Plants and trees are used as building materials.

- The Indians have a use for 90 per cent of forest plants.

- Face and body paints are made from plants.

- Body ornaments are made from animal bones and teeth, birds feathers and tree bark.

- Nine species of stingless bee are used to produce honey.

- Musical instruments are made from animal bones and skins, bamboo and nutshells.

- Medicines are made from over 650 different forest plants.

Kayapo Indians in the forest

Forest care

- Small-scale farming allows soil to be used again later.

- Limits are put on number of wild animals and birds hunted to ensure species survival.

- Clean and orderly villages help reduce waste and pollution.

- Useful new trees and shrubs have been added to the forest.

- Soil quality is improved in the clearings by adding fertiliser made from plant leaves, ashes and termite nests.

- Spare food and forest products are traded to reduce waste.

Activities

1 Answer the questions below. The first letters are given in drawing **E**. Write out your completed answers.

a) Least populated region of Brazil
b) An Amerindian tribe
c) Where the Kayapo live
d) Worn by Kayapo men
e) A forest clearing used for growing crops
f) A type of farming, where people move from one place to another
g) Where just enough food is grown for the village
h) Someone who collects and hunts food
i) A method of forest clearance

E

2 a) Make a larger copy of diagram **F**.
b) Put these labels into the correct boxes.

Cut down the forest

Plant crops

Soil loses goodness

Burn the vegetation

Harvest crops

Abandon clearing and move on

c) Why is clearing the forest called slash and burn?
d) Why is this method of farming called shifting cultivation?

Shifting cultivation

F

3 Draw a star diagram to show how useful the forest is to the Kayapo Indian. Write no more than four words for each use.

4 a) How do the Indians reduce waste in the forest?
b) How do the Indians care for the forest?

Summary

The Indians depend on the forest for their life. They have a great respect for all plants and animals and are careful not to damage the environment.

What is happening to the Amazon rainforest?

The Amerindians are not the only people with an interest in the rainforest. Since the 1970s the Brazilian government has encouraged the development of the Amazon region. The aim has been to bring wealth to the area by using its natural resources. Some of the developments are shown below.

A

Mining Huge deposits of iron ore, gold, copper, bauxite and other minerals have been discovered in the rainforest. Mining companies have felled trees and built roads through the forest to reach these deposits.

B

New roads Many new roads have been built deep into the rainforest. The Trans-Amazonian Highway is the longest. It stretches 5300 km (3300 miles) across Brazil from east to west. The roads help move timber, cattle and crops to markets. People from other parts of Brazil have settled along these roads.

C

Power stations An unlimited water supply and ideal river conditions have led to the development of many hydro-electric power (HEP) stations. HEP stations provide cheap and plentiful energy for industry, transport and domestic use. More than 125 new HEP dams are to be built in the next 15 years. The reservoirs behind the dams flood large areas of forest.

D

Ranching Large areas of the forest have been bought by multi-national companies for cattle ranching. These companies have burnt down the forest and replaced the trees with grass. The beef from the cattle ranches has largely gone to fast-food chains in the United States and Europe to make into burgers.

Developments in the rainforest have brought many benefits. They have provided jobs for people, brought money into the region and helped Brazil's industry and agriculture. However, there have been problems. Large parts of the rainforest have been destroyed and the Amerindian way of life has been put under serious threat.

6000 people lost their homes because of the Tucurui HEP scheme

New roads have brought in settlers. Some of our people have been murdered by them.

Cattle ranchers burn the forest, ruin the soil and leave the mess for us

New roads cut through the forest and damage our land

Some of the plants we use to make medicines have disappeared

We don't want development, we just want to live as we are

Mining waste and chemicals have poisoned our rivers

Huge areas of forest have been cleared for mining

There is not as much wildlife now that the trees have been felled

We were forced off our land by violent cattle ranchers

Many Indians have died from diseases brought in by settlers

HEP schemes have put large areas of forest under water

E

Chief Raoni of the Kayapo

❛ We want nothing from the white man. He has brought us only death, illness and murder. He has stolen and destroyed our forest. We want to be left alone to live with our ancestors. We want to stay here. It is our right. ❜

Activities

1 a) What minerals may be found in the forest?
   b) What is the cattle ranching meat mainly used for?
   c) What advantages have new roads brought to the area?
   d) Why is the Amazon region good for HEP?

2 Give three benefits that developments in the rainforest have brought.

3 Look at photo **A** of mining in the rainforest. For each of the following people say what their feelings might be about the new development. Give reasons for your answers.
   ● A Kayapo Indian
   ● A Brazilian mine owner
   ● A European tourist

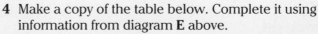

4 Make a copy of the table below. Complete it using information from diagram **E** above.

Problems of development				
Mining	New roads	HEP	Ranching	Others

5 Work with a partner for this task.
   a) Imagine you are Chief Raoni. Make a list of reasons why you are against rainforest development.
   b) Imagine you are a government minister. Make a list of reasons why development of the rainforest should go ahead.
   c) Display your arguments as an attractive and interesting poster.

Summary

There have been many changes to the Amazon rainforest. These have brought benefits to some people but caused problems for others.

Two regions compared

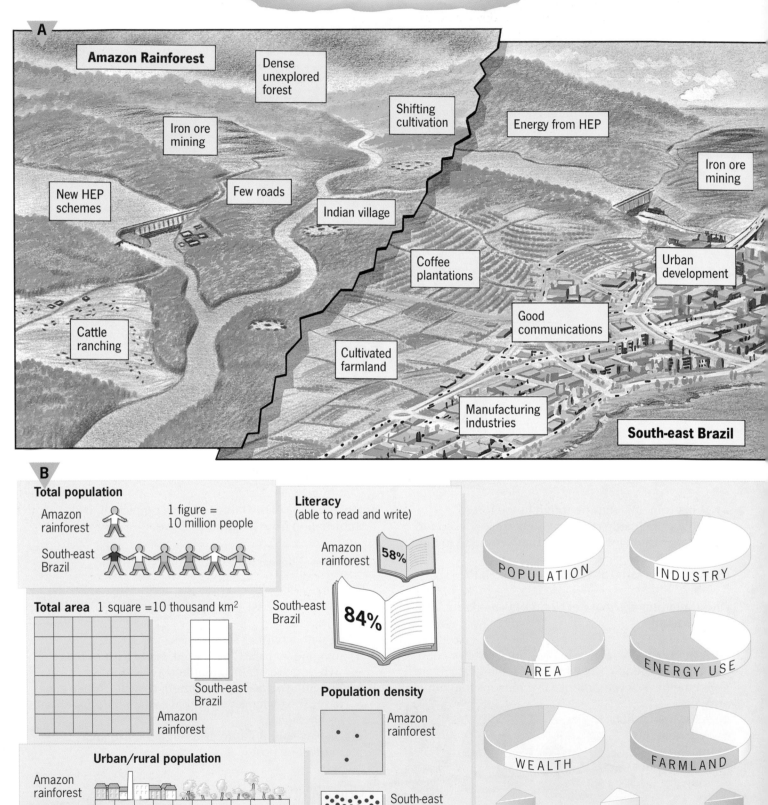

A

Amazon Rainforest

Dense unexplored forest

Shifting cultivation

Energy from HEP

Iron ore mining

Iron ore mining

New HEP schemes

Few roads

Indian village

Coffee plantations

Urban development

Cattle ranching

Cultivated farmland

Good communications

Manufacturing industries

South-east Brazil

B

Total population

Amazon rainforest

1 figure = 10 million people

South-east Brazil

Total area 1 square = 10 thousand km²

South-east Brazil

Amazon rainforest

Urban/rural population

Amazon rainforest

100

South-east Brazil

100

Literacy
(able to read and write)

Amazon rainforest **58%**

South-east Brazil **84%**

Population density

Amazon rainforest

South-east Brazil

• = person per km²

POPULATION

INDUSTRY

AREA

ENERGY USE

WEALTH

FARMLAND

Amazon rainforest

South-east Brazil

Rest of Brazil

As we have seen, South-east Brazil is very different to the Amazon rainforest (diagrams **A** and **B**). The South-east is a small region but it has nearly half of Brazil's population and most of its wealth. The area's original wealth came from enormous coffee plantations. Nowadays, modern industry powered by cheap hydro-electricity is the biggest source of income. South-east Brazil is densely populated. It has the country's largest cities, densest network of roads and railways, and highest standards of living.

The Amazon rainforest is in complete contrast. The area is huge yet very few people live there. Most of the population is concentrated in river towns like Manaus. The Amerindians live mainly in isolated villages deep in the jungle. The region is poor and provides only a tiny proportion of Brazil's wealth.

Although this area is rich in resources, development is proving difficult because of unhealthy living conditions, poor communications and vast distances. There is also concern that developments can damage the rainforest and destroy the Indian way of life.

Activities

1 Draw table **C** and complete it by putting the features from drawing **A** in the correct columns.

C

Amazon rainforest	South-east Brazil	Amazon rainforest and South-east Brazil

2 a) Make two copies of star diagram **E** below.
  b) In the centre of the first one write 'Amazon rainforest'. In the centre of the second one write 'South-east Brazil'.
  c) Unscramble the words below and use them to complete your diagrams. Some words can be used more than once.

D

PooaSula
rooeinr
ewt
toh
wol
ehahlty
slepaatn
ertopuseug
aprsysle
eedsnyl
eHolzoBritnoe
oreiRdjieoan
gihh
toh

E

1 The main language is _____

2 The climate is _____ and _____

3 The area is _____ populated

4 The main town or towns _____

5 Most people have a _____ standard of living

6 A mineral mined here is _____

F

Fact File

	Amazon rainforest	South-east Brazil
Population (most or least)		
Area (largest or smallest)		
How crowded (most or least)		
Urban population (most or least)		
Able to read or write (most or least)		
% of Brazil's industry (57 or 3)		
% of Brazils wealth (4 or 57)		
% of Brazil's cultivated farmland (33 or 2)		

3 a) Make a larger copy of Fact File **F** above.
  b) Choose from the brackets the correct word or number for each column.
  c) Add more information to your Fact File to help you compare the two regions.

4 Imagine that you are a writer for a geographical magazine. Write two short articles of about 100 words each to describe the main features of:
  a) the Amazon rainforest
  b) South-east Brazil.

Summary

The South-east and the Amazon rainforest are two important but very different regions of Brazil.

How developed is Brazil?

A

The good life – Copacabana beach, Rio de Janeiro

Carajas iron ore mine doubles production

A poor family in a Rio de Janeiro favela

New HEP station opens at Tucurui

HIGH-TECH SPACE PROGRAMME READY FOR SATELLITE LAUNCH

Modern office blocks, São Paulo

An old forge in the poor part of town

Health-carers helpless as killer disease strikes Indians

CAR INDUSTRY ACCLAIMED AS SUCCESS OF THE DECADE

FOOD SHORTAGES A PROBLEM FOR CITY HOMELESS

Niteroi bridge, Rio de Janeiro

Horse drawn water cart near Manaus

Rio – over two million slum dwellers

Young Brazilians top unemployment league

As we have seen, countries are at different levels of **development** or progress. **Developed countries** are generally wealthy. They have high levels of industrial development and most of the people have a good standard of living. **Developing countries** are usually poor. They are in the process of building up their industry and farming. For many of their people the benefits of good health care, education and housing may not be available. Some people in these countries do not even have enough food to eat.

So how developed is Brazil? Look at the information above and try to decide.

As you will see from the previous page it is really quite difficult to label Brazil as either developed or developing. There are many people in the country who are wealthy and who have a very good standard of living. There are many others, however, who live in poverty and do not have adequate food, housing or health care. This is the problem in Brazil. The rich are very rich but the poor are very poor.

In fact Brazil is a rapidly developing country. Since the 1960s it has made tremendous industrial progress and is now probably the ninth wealthiest country in the world. As yet, however, the wealth has not reached many of Brazil's people and for them, hardship continues to be a way of life.

Like many less developed countries, Brazil has had to make choices about the way to develop. It has chosen rapid industrialisation as its way forward. This has brought wealth and economic success to the country. Unfortunately for many of its people, it has not yet brought an improvement in standards of living.

Look at the table B. It compares Brazil with some of her South American neighbours and the UK using **development indicators**.

GNP and energy use are examples of **economic indicators**. They tell us about the general wealth of the country.

Social indicators like infant mortality and school attendance help us measure the standard of living that people have.

B

Indicator of development	Argentina	Brazil	Bolivia	Colombia	Ecuador	Paraguay	Peru	UK
GNP per capita (US$) (earnings per person per year)	8030	3640	800	1910	1390	1690	2310	18 700
Energy use (tonnes of coal equivalent per 1000 people)	183	77	36	77	73	23	49	499
Adult literacy (%)	95	81	78	87	86	90	85	99
Infant mortality (deaths per 1000 births)	29	57	93	37	57	39	76	8

Activities

1 Give the meaning of the terms:
   ● Development
   ● Developed country
   ● Developing country.

2 Look at the information on the opposite page. Make a list of all the features you would expect to find in:
   a) a developed country;
   b) a developing country.
   c) Why is it difficult to describe Brazil as either developed or developing?

3 List the South American countries from table **B** in order of:
   a) GNP per capita (highest first)
   b) Energy use (highest first)
   c) Secondary school attendance (highest first)
   d) Infant mortality (lowest first)

4 Copy out and complete the following sentences.

   ● Using economic indicators, Brazil is ____ developed than most South American countries.
   ● Using social indicators, Brazil is ____ developed than most South American countries.
   ● Using both economic and social indicators, Brazil is ____ developed than the UK.

5 Brazil has been described as 'a rich country full of poor people'. Suggest reasons for this description.

Summary

Brazil has developed quickly since the 1960s. It is now one of the world's leading industrial countries. Many Brazilians have gained little from this development.

How interdependent is Brazil?

Goods sold to other countries are called **exports**

The money earned from exports can help pay for **imports**

Goods that are brought into a country are called **imports**

All countries need help from other countries if they are to make progress. Help for Brazil has come in many ways. For example:

● Foreign banks have given loans;
● Large **multi-national companies** have invested money in developing resources and industry;
● Rich countries and voluntary organisations have provided **aid**;
● **Trade** has been developed with countries around the world.

When countries work together like this, and rely on each other for help, they are said to be **interdependent**.

The development of trade has been particularly important to Brazil. In the past, **primary goods** such as coffee and soya beans have been the main **exports**. Recently, with the rapid growth of industry, more money has been made from exporting **manufactured products** such as cars, clothes and shoes. Brazil also needs to **import** goods and resources from other countries (diagram **A**). These include oil, chemicals and heavy machinery for industry. A large variety of foodstuffs that are not available in Brazil are also imported.

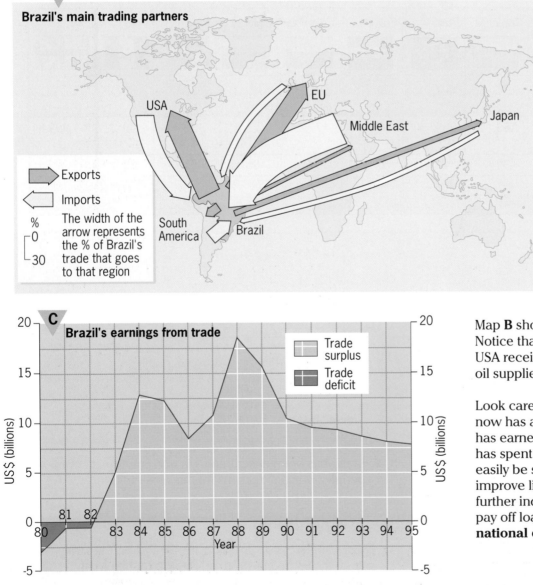

B Brazil's main trading partners

USA

EU

Middle East

Japan

South America

Brazil

➡ Exports

⬅ Imports

% The width of the arrow represents the % of Brazil's trade that goes to that region
0
30

C Brazil's earnings from trade

US$ (billions)

Trade surplus

Trade deficit

81 82

80 83 84 85 86 87 88 89 90 91 92 93 94 95

Year

Map **B** shows Brazil's main trading partners. Notice that the **European Union (EU)** and the USA receive over half of Brazil's exports. Most oil supplies are imported from the Middle East.

Look carefully at graph **C**. It shows that Brazil now has a **trading surplus**. This means that it has earned more money from exports than it has spent on imports. The extra money can easily be spent. It can be used to improve living standards or be invested in further industrial development. It can also help pay off loans and so reduce Brazil's huge **national debt**.

In the past, political problems have restricted Brazil's interdependence with other South American countries. Now however, these problems have largely been overcome. In the last ten years, the biggest increase in Brazil's foreign trade has been with other countries on the South American continent.

Many countries around the world have joined together to form trading organisations. These organisations encourage trade between countries and also work together and help each other in many other ways. In doing this they become interdependent.

Brazil is a member of several groups. One of these is LAFTA, the Latin American Free Trade Association of which Brazil has been a member for many years. LAFTA's membership includes Mexico as well as most of the South American countries. Another is MERCOSUL, a small group of four countries which have recently joined together as the South American Common Market. The aims of MERCOSUL are shown in diagram **D**.

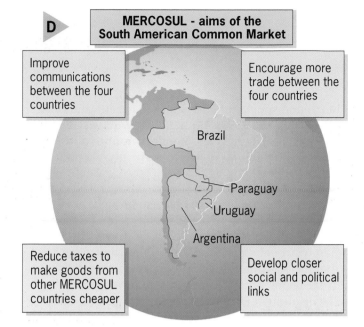

D ▶ MERCOSUL - aims of the South American Common Market

Improve communications between the four countries

Encourage more trade between the four countries

Reduce taxes to make goods from other MERCOSUL countries cheaper

Develop closer social and political links

Activities

1 Make a copy of crossword **E** and complete it using these clues.
   1 Export products like cars, clothes and shoes
   2 Money owed by a country
   3 The exchange of goods and materials between countries.
   4 A large business or company
   5 A trade organisation
   6 This is a form of help
   7 Goods or resources brought into a country
   8 A trading organisation
   9 When countries work together and rely on each other.
   10 Goods or resources sold to another country
   11 A major primary export for Brazil

2 a) Make a copy of table **F** and complete it with help from map **B**.
   b) List Brazil's export markets in order of importance. Give the largest first.
   c) List Brazil's import markets in order of size. Give the largest first.

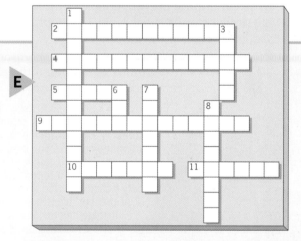

E ▶

3 a) Why is there a high percentage of imports from the Middle East?
   b) Why might you expect the percentage of exports to other South American countries to be higher? (Think of Brazil's location.)

4 a) What is a trading surplus?
   b) In which year was Brazil's first trading surplus?
   c) In which year was Brazils highest trading surplus and how much was it worth?
   d) What can Brazil do with a trading surplus?

5 a) What is meant when a country is said to be interdependent?
   b) Give four ways in which a country may be interdependent.

Name of region	Imports (%)	Exports (%)	
	20	24	**F**
	5	30	
	15	10	
	44	4	
	6	7	

Summary

The development of links with countries in South America and the rest of the world have helped Brazil to become interdependent.

Can development be sustainable?

Sustain: to keep in existence; to keep up; to maintain, prolong

Brazil has made spectacular progress in recent years. During the 1960s and 1970s industrial growth was so rapid and successful that it was called an **'economic miracle'**. The miracle brought great riches to some people and made Brazil very wealthy.

Unfortunately industrial growth has also brought problems. As we have seen, large areas of the Amazon rainforest have been cleared, wildlife has been lost and the Indian way of life has been threatened. Some people worry that this so-called development and progress is actually damaging Brazil . They say that development should bring about an improvement in living conditions for people. At the same time, however, it should not harm or destroy the environment nor prevent people in the future from achieving similar standards of living. Development like this is called **sustainable development**.

Sustainable development is sensible development. It uses but does not waste resources. It improves but does not threaten ways of life. It looks after the needs of today but does not damage the future.

Sustainable development needs good planning. It also needs the co-operation of people and nations and it requires a commitment to conservation. Brazil and other countries in the world are now showing a greater concern for the damaging effects of industrial growth. Steps to improve the quality of development are already being made. We just hope that it's not too late!

Commercial logging is one example of an industry that if managed correctly may be sustainable.

Commercial logging – the problems

Commercial logging is the second largest cause of rainforest destruction. Most of the timber taken from the forest is exported to the richer countries.It is used in the construction industry and for making furniture, plywood, veneers, wood pulp and paper. Trees are a renewable resource, but only if they are used and managed carefully. At present this is not happening and logging has become a wasteful and damaging forest activity.

- After forest clearance the soil quickly loses its fertility and is unable to support new vegetation

- The areas of land that have been cleared are so large that it is difficult for vegetation to return

- Because of costs, logging companies rarely replace felled trees with new ones

- Amazon Indians have been moved off their land and wildlife has been destroyed

- Government regulations are rarely heeded

- Heavy machinery compacts the soil and damages drainage. This makes recultivation difficult.

- Only about 5% of the forest trees are actually wanted by the loggers. Unfortunately to obtain those, they damage or destroy 65% of the remaining ones.

- Soil left without tree cover is quickly washed away by heavy rains. Insufficient soil is left for new plant growth.

Commercial logging – a sustainable industry?

We are considering whether to put a tax on hardwood products. The money collected would be spent on reafforestation schemes.

We have created areas where logging activities are either banned completely or strictly controlled

National Park Service of Brazil

The International Trade and Timber Organisation

The United Nations

Our intention is to reduce international trade in logging. We think that this can be done by taxing endangered hardwoods such as mahogany and so reduce logging profits.

The International Tropical Timber Organisation

Our main advice to logging companies is to:
- limit the number of trees felled in an area
- leave enough trees for the forest to recover
- protect the trees that are left
- plant two trees for every one tree felled
- have a working plan that is agreed by conservationists and is enforced.

Manager of Habitat shop

Like many other high street furniture shops, we have banned the sale of all rainforest products that are not produced in a sustainable way.

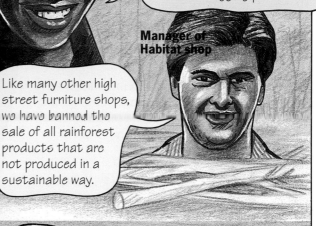

We only want certain trees from the forest so we have cleared some areas and planted only the types that we need. This makes it much easier to look after the trees and means that we don't have to keep destroying more and more forest.

Plantation owner

Worldwide Fund for Nature

We recognise the need for timber products. Our researchers are trying to find out how many trees may be felled in an area without causing permanent damage to the wildlife.

Activities

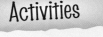

1 a) Write out the meaning of 'sustain'.
   b) Give four facts about sustainable development.
   c) Give three needs of sustainable development.

2 a) Make a list of at least ten uses of timber.
   b) Give three ways in which commercial logging can damage the soil.
   c) Why is logging a wasteful and damaging activity?

3 One way of making the logging industry more sustainable is to reduce the amount of timber taken from the forest.
   a) Which organisations support this method?
   b) How do they plan to bring this about?

4 Imagine that you work for the International Tropical Timber Organisation. Design a poster to show how you think logging can be made a sustainable industry. Try to add drawings for each of your suggestions.

Summary

Sustainable development is progress that can go on year after year. If practised properly, logging in the rainforest could be a sustainable industry.

3 Kenya

What is Kenya like?

Kenya is located on the Equator where it crosses central Africa. Photograph **A** shows Africa taken from space. On it Kenya looks quite a small country but in fact it is two and a half times **bigger** in area than the UK. However, its total population is two and a half times **smaller** than that of the UK.

Although Kenya is, economically, one of the poorer countries in the world, in terms of scenery and wildlife it is one of the richest. Its spectacular scenery includes vast lakes, high mountains, numerous volcanoes, huge stretches of sandy beaches, and even a snow-capped peak. Its natural vegetation includes desert, grassland and tropical trees. Kenya has a wealth of wildlife which includes many of the world's largest, and some of the rarest, animals and birds. It also has a population which has considerable differences in colour, religion and language.

What, if anything, do you already know about Kenya? Some of you may have seen Kenyan long distance runners winning many of the world's top athletic events (photo **B**). Others may have heard of the Maasai (photo **C**). Many more of you are likely to have watched travel programmes about Kenya on television. These programmes usually show holiday makers either relaxing in large, modern coastal hotels (photo **D**) or taking a wildlife safari (photo **E**).

A Satellite photo of Central Africa

Sahara Desert

Equator (0°)

Kenya

Atlantic Ocean

Lake Victoria

Nairobi

Indian Ocean

N

B Another gold medal for Kenya

C Maasai in their traditional red cloaks

38

D

E

Activities

1 Work with a partner and make a list of things which come to mind when you think about Kenya.

2 a) Using clues on these pages, name:
- the continent in which Kenya is located
- the line of latitude which passes through Kenya
- the largest lake in Kenya
- Kenya's capital city
- the ocean to the east of Kenya
- a holiday on the coast
- a wildlife holiday
- a group of people who herd animals in Kenya
- a very large grass eating animal found in Kenya
- a large animal which kills other animals for its food.

b) Fit your answers into a copy of figure **F**.

F

E

A

I

Summary

Although Kenya is not a rich country, it does have a wealth of beautiful and contrasting scenery and wildlife.

Physical features in Kenya

Kenya is part of a large continental **plate** that was once a flat, barren plain. This plate, like other plates, moved slowly across the earth's surface. In time it **collided** with a neighbouring plate. This collision pushed the land upwards causing it to buckle – like when a car crashes into a brick wall. One result of this buckling was the formation of huge cracks in the earth's surface. Two of the largest cracks ran side by side, not just through Kenya, but almost the whole length of Africa.

In Kenya, the land between the cracks collapsed to leave a very deep, steep-sided **rift valley** (photo **A**). Molten rock, known as the **magma**, was able to force its way upwards through the cracks to the earth's surface. Where it reached the surface, as **ash** or **lava**, it formed large volcanoes. These volcanoes have formed a line of mountains on either side of the rift valley They have also produced smaller **cones** on the flat valley floor (photo **B**). Later, water flowing into the valley was trapped and formed numerous lakes.

The Rift Valley in Kenya **A**

Volcano with secondary cone

B

Diagram **C** is a section drawn across Kenya to show the position of the rift valley and the landforms linked with it. Notice the old volcano of Mount Kenya. Even although it lies on the Equator it is high enough to have snow lying on its summit all year round (photo **D**).

Diagram **C** also shows how the land gradually slopes downwards to the sea. Along almost the whole length of Kenya's coast there are **sandy beaches** and a **coral reef** (photo **E**). The reef is one kilometre from the shore and protects the beaches from large storm waves. The only gap in the reef large enough for ships to sail through is opposite the port of Mombasa.

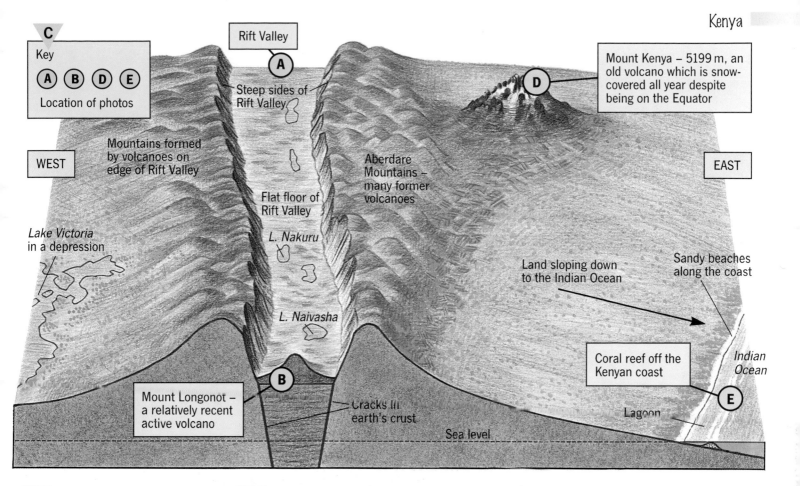

C

Key

Ⓐ Ⓑ Ⓓ Ⓔ

Location of photos

WEST

Rift Valley

Ⓐ

Steep sides of Rift Valley

Mountains formed by volcanoes on edge of Rift Valley

Flat floor of Rift Valley

L. Nakuru

Lake Victoria in a depression

L. Naivasha

Mount Longonot – a relatively recent active volcano

Ⓑ

Cracks in earth's crust

Sea level

Aberdare Mountains – many former volcanoes

Mount Kenya – 5199 m, an old volcano which is snow-covered all year despite being on the Equator

Ⓓ

EAST

Land sloping down to the Indian Ocean

Sandy beaches along the coast

Coral reef off the Kenyan coast

Indian Ocean

Ⓔ

Lagoon

D Mount Kenya

E Coral reef off the Kenyan coast

Activities

1 Make a copy of diagram **F**.
   a) Label two places with:
      ● mountains formed by volcanoes
      ● lakes
      ● cracks in the earth's crust.
   b) Label one place where there is:
      ● a rift valley
      ● a coral reef.

2 a) What is a rift valley? How does a rift valley form?
   b) Why are volcanoes found along a rift valley?
   c) Why do lakes form on the floor of a rift valley?

F

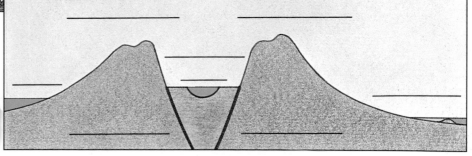

Summary

Many of Kenya's most attractive landforms can be seen along either the rift valley or the coast.

People in Kenya

The earliest of man's ancestors lived in parts of the rift valley in Kenya. More recently, people have moved into the country from several directions. This has resulted in many groups of people living together and sharing their different ways of life.

The movement of people into Kenya

Over 98 per cent of the present inhabitants of Kenya are Africans. They come from over 30 different **ethnic groups**. Each group has its own language and way of life. They arrived in Kenya from three different directions (map **A**). Although most Kenyans still identify themselves with one specific group, their way of life has been changed through marriages and contact with people from other groups.

Most of the remaining 2 per cent have come from Asia. They are usually either Arabs or Indians. The Arabs tend to live along the coast while most Indians live in Nairobi. A small group of Europeans, mainly British, also live in Kenya.

Where do people live in Kenya?

The population of Kenya, as in most other countries, is not evenly spread out. Some places are very crowded while others have very few people living there. This spread, or **distribution**, of people is shown on map **A**. The map also gives some of the reasons for the uneven distribution.

How is Kenya's population changing?

The total population of Kenya is increasing very rapidly (graph **B**).

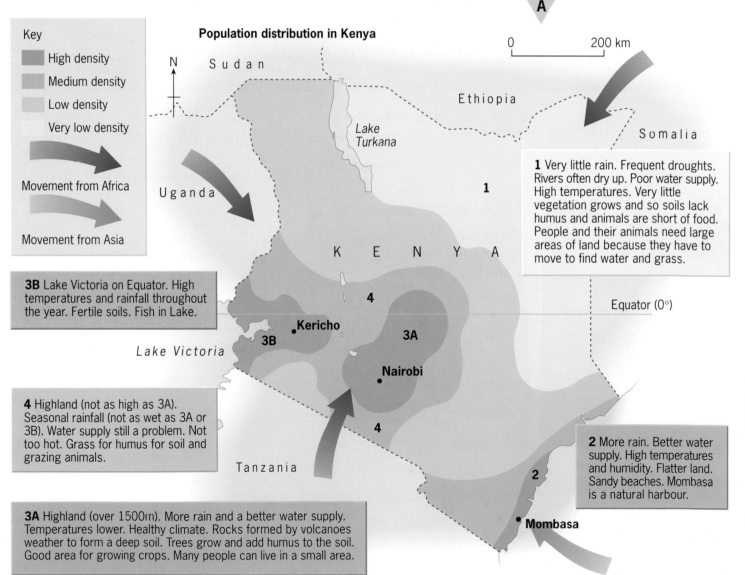

A

Population distribution in Kenya

0   200 km

Key
- High density
- Medium density
- Low density
- Very low density

Movement from Africa

Movement from Asia

1 Very little rain. Frequent droughts. Rivers often dry up. Poor water supply. High temperatures. Very little vegetation grows and so soils lack humus and animals are short of food. People and their animals need large areas of land because they have to move to find water and grass.

3B Lake Victoria on Equator. High temperatures and rainfall throughout the year. Fertile soils. Fish in Lake.

4 Highland (not as high as 3A). Seasonal rainfall (not as wet as 3A or 3B). Water supply still a problem. Not too hot. Grass for humus for soil and grazing animals.

3A Highland (over 1500m). More rain and a better water supply. Temperatures lower. Healthy climate. Rocks formed by volcanoes weather to form a deep soil. Trees grow and add humus to the soil. Good area for growing crops. Many people can live in a small area.

2 More rain. Better water supply. High temperatures and humidity. Flatter land. Sandy beaches. Mombasa is a natural harbour.

Sudan   Ethiopia   Somalia   Uganda   Lake Turkana   K E N Y A   Kericho   Lake Victoria   Nairobi   Tanzania   Mombasa   Equator (0°)

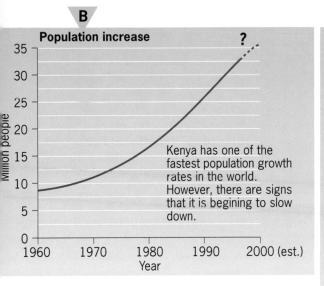

B Population increase **?**

Kenya has one of the fastest population growth rates in the world. However, there are signs that it is begining to slow down.

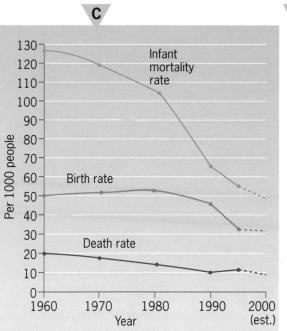

C

D

The life expectancy for a Kenyan man was 39 in 1960. It is likely to be 65 in 2000

The life expectancy for a Kenyan woman was 43 in 1960. It will probably to be 70 in 2000

In fact Kenya's population doubled in size between 1970 and 1990. Graph **C** shows that this increase was mainly due to

- a very high **birth rate** (one of the highest in the world)
- a falling **death rate** (graph **C**).

When birth rates are higher than death rates there is a **natural increase** in population.

Other reasons why Kenya's population is likely to continue to grow include:
- The rapid decline in **infant mortality** which means that fewer

babies die in the first year of their lives (graph **C**)
- Half of Kenya's population are aged 15 or under. This suggests that in a few years there will be even more people of child-bearing age.
- At present only a small proportion of the population is aged over 60. However, as life expectancy steadily increases, then more of

the population are likely to live longer (diagram **D**).

The rapid growth of Kenya's population means that the country's raw materials and natural resources have to be shared among a greater number of people.

Activities

1. a) Which of the regions numbered on map **E** is the least crowded?
   b) Give three reasons why few people live in this region.
   c) Give three reasons why many people live along the coast.
   d) Which of the regions numbered on map **E** is the most crowded?
   e) Give three reasons why so many people live in this region.

2. Make a copy of table **F**. Use the information given in diagrams **B**, **C** and **D** to complete table **F**.

E

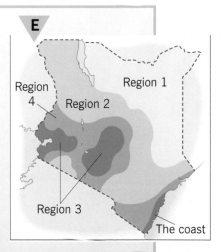

Region 1

Region 4

Region 2

Region 3

The coast

F

	1960	2000	Trend (rising or falling)
Total population	(million)	(million)	
Birth rate	(per 1000)	(per 1000)	
Death rate	(per 1000)	(per 1000)	
Infant mortality rate	(per 1000)	(per 1000)	
Life expectancy — Males			
Life expectancy — Females			

Summary

Kenya's population is very uneven in its distribution and is one of the fastest growing in the world.

43

Holidays in Kenya

There are few countries in the world that can offer the tourist the variety of scenery and wildlife that Kenya can. Inland there are mountains, lakes, grassy plains and a wealth of wildlife. On the coast there are sandy beaches and coral reefs. Tourism has become Kenya's fastest growing industry and largest earner of money from overseas.

To encourage tourism and to protect its scenery and wildlife, Kenya has set up over 50 **National Parks** and **game reserves**. Many visitors choose to visit some of these parks on organised tours known as '**safaris**' (advert A). They are taken around in minibuses by drivers who are also expert local guides. The 7- or 9-seater minibuses have roofs and windows which open for viewing and photography.

A

Safari

Day 1
Nairobi/Samburu (310 km)
After breakfast drive north, cross the Equator and pass Mt Kenya, to Samburu Lodge. After lunch there will be a game drive when you should see elephant, buffalo, lion, giraffe, zebra, crocodile and many species of bird. Overnight at Samburu Serena Lodge right on the river.

Day 2
Samburu
Early morning game drive. Relax at mid-day around the swimming pool or watch the Samburu perform traditional dances. Late afternoon game drive. Overnight at Samburu Serena Lodge.

Day 3
Samburu/Lake Nakuru (200 km)
After early morning game drive, drive to the Thomson's Falls, and down into the Rift Valley to Nakuru for lunch. An afternoon drive will let you see vast flocks of flamingos and the endangered Rothschild giraffe. Overnight at Nakuru Lodge.

Day 4
Lake Nakuru/Maasai Mara (340 km)
Drive through the Rift Valley passing Lake Naivasha on the way to the Maasai Mara game reserve. Afternoon game drive. Overnight at Keekorok Lodge.

Day 5
Maasai Mara
The huge Mara plain provides some of the best game-viewing in East Africa. During early morning and late afternoon game drives you are likely to see huge herds of wildebeest and zebra, as well as lion, elephant, cheetah, leopard, giraffe, and hippo. An option is the early morning Balloon Safari. Overnight at the Keekorok Lodge.

Day 6
Maasai Mara/Nairobi (260 km)
Early morning departure arriving at Nairobi for lunch. Afternoon flight to Mombasa to continue your holiday at a beach hotel.

Map:
```
0        200 km          Lakes/Oceans
                         National Parks/
                         Game Reserves
                      ▲  Mountain peaks
                         National boundary
N                     ←  Route
↑
       K E N Y A
                      Samburu National
Lake                  Reserve
Victoria  Mt. Kenya
          5200m ▲        Equator (0°)
   Lake
   Nakuru
          Nairobi
Maasai
Mara National
Reserve
                              Indian
Amboseli                      Ocean
National Park
   Mt. Kilimanjaro
   5895m              Mombasa
```

Not every visitor to Kenya goes on safari. Many tourists take advantage of the coastal climate. Because Kenya lies on the Equator, the climate is usually hot and sunny throughout the year. These visitors, who are often joined by weary safari travellers, prefer to relax at one of many beach resorts (advert **B**). The resorts have clean, sandy beaches. The warm sea, protected from storms by the coral reef, is ideal for water sports. The beach resorts extend along most of the coast either side of Mombasa.

B ▼

Nyali Beach

*O*ne of Kenya's best known hotels is managed by Block Hotels and is situated four miles north of Mombasa in delightful grounds leading directly to a palm-fringed beach.

A good choice of restaurants and bars and regular evening entertainment and daytime activities are arranged. There are three swimming pools, including one for children. Nyali also boasts a Jacuzzi, snooker table, shops, hairdresser and beauty salon. Children have their own club with a supervised activities programme which leaves you free to be as lazy or as active as you choose. There is free use of pedaloes, canoes and minisails. Other watersports include waterskiing, windsurfing, sailing, scuba diving and glass bottom-boat rides.

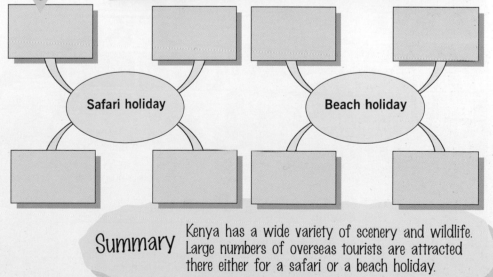

Activities

C ▼

1 Draw two star diagrams, like the ones on the right, to show at least four attractions of a safari holiday and four attractions of a beach holiday.

2 Design a poster for Kenya's Tourist Office. The poster should show the attractions of **either**
a) a 'Safari holiday', **or**
b) a 'beach resort holiday'.
Try to include a map, a labelled photo or a collage of several photos, and a short written description.

Safari holiday

Beach holiday

Summary Kenya has a wide variety of scenery and wildlife. Large numbers of overseas tourists are attracted there either for a safari or a beach holiday.

The benefits and problems of tourism

Kenya realises the importance of its spectacular scenery and wildlife as a major tourist attraction. The growth of tourism can bring many benefits to a country such as Kenya (diagram **A**). It can also create many problems. Large numbers of people now go on **safaris** or stay at **beach resorts** (diagrams **B** and **C**). However, their numbers can damage the very environments which attracted them there in the first place. Tourism puts pressure upon **fragile environments**, wildlife and local people.

A Advantages of tourism

Tourism can create lots of new jobs. We are very short of work.

Money from tourism has helped to improve our roads and to build more houses

Tourism brings a lot of overseas money into Kenya

Farmers have a larger market for their produce

We can make souvenirs to sell to the tourists

I hope to become a tourist guide or a safari driver when I leave school

Money from tourism can be used to build schools and hospitals

B Problems created by safaris

The environment

Minibuses are meant to keep to well defined tracks but drivers often take short cuts to get as near to the animals as possible. This can cause tracks to be widened, to turn dusty in the dry season and marshy in the short rainy season. Both increase soil erosion.

Wildlife

Minibuses should not go within 25 metres of the animals. Drivers often ignore this rule as their passengers are more likely to give them good tips if they get very close to the animals. This may prevent the animals from feeding, drinking and mating.

Hot air balloons have gas burners which make a loud noise from time to time. Conservationists think this, and the balloons' moving shadows, can also disturb wildlife.

People

Apart from people working at safari lodges, nobody can live in the National Parks. We were forced to leave our homes and traditional grazing grounds when the National Park was opened. Most of the money which tourists spend goes to the government, not to us local people.

C Problems created by beach holidays

Farm land is lost – Some farmers lose their jobs and homes and the food supply is reduced

Busy coastal road due to tourist traffic

Natural vegetation cleared for building expensive hotels and beach resorts

The coral reef – a fragile environment – is exposed at low tide

Wildlife disappears when its natural habitat is destroyed

Beaches and sheltered lagoons may be polluted

Coral is killed
- by boat anchors
- by tourists trampling over it
- by sand disturbed by snorkellers (suffocation)
- if broken off for souvenirs

Activities

1 With the help of diagrams **A** and **D**:
a) Give two reasons why an increasing number of tourists visit Kenya.
b) Give two pieces of evidence to show that this increase has been rapid.
c) Give three ways by which tourism can help local employment.
d) Give three uses of the income (money) received from tourists.

2 a) What is meant by the term 'fragile environment?'
b) Make a copy of table **E**. Give it the title, 'Problems created by safari and beach holidays'. List the problems under these headings:
- environment
- wildlife
- local people.

3 On balance, do you think that tourism is a good or a bad thing for Kenya? Write a short paragraph to explain your answer.

D

	Tourist beds available (hotels and lodges)	Overseas visitors	Income from tourism (£ Kenyan)	
1965	2,100,000	224,000	127 million	= 6%
1986	9,371,000	465,000	247 million	= 12% overseas money
1992	12,037,000	560,000	594 million	= 20%

E

	Problems	
	Safari holidays	*Beach holidays*
Environment		
Wildlife		
Local people		

Summary

Tourism can bring jobs and earn money for countries such as Kenya. However, it can also cause problems for people and wildlife and can harm the environment.

Urban life in the Nairobi region

Nairobi is the capital city of Kenya. As in many large cities in developing countries the people who live and work in Nairobi are usually either well-off or very poor. The gap between the few who are rich and the majority who are poor is much larger than in economically more developed countries. In other words the rich are very rich and the poor are very poor.

Kip and Krista come from well-off families. They live, like most other well-off people, near to the city centre. Their parents have jobs which are full time and well paid. The best jobs are in the city centre, especially in various government buildings, tall office blocks and banks (photo A). Kip and Krista can afford the cost of travel into the city centre. They often make this journey to visit either the more expensive shops or the best places of entertainment.

Their houses are large and contain modern amenities such as air-conditioning, electricity and a bathroom with a shower. Each family has its own car, TV and video, and many modern gadgets. However, the houses need to be protected by good security systems. Kip and Krista both attend nearby private schools and hope eventually to go to Nairobi University.

Nairobi – the city centre **A**

- Location of Mathare Valley (shanty settlement)
- Main shopping area with hotels, restaurants and cinema
- Main banks and financial buildings
- Kenyatta International Conference Centre
- Parliament buildings
- Inner ring road
- Uhuru (Freedom) Park and sports arena

B

Activities

1 Make a list of the features found in the centre of Nairobi that:
   a) may also be found in your local town or nearest city centre;
   b) would not be found in your local town or nearest city centre.

2 Diagram **B** is a simplified map of Nairobi. Make a copy of the map and put these labels in their correct place:
   - large houses with many amenities;
   - shops and offices;
   - shanty settlements;
   - small houses with very few amenities;
   - well paid jobs;
   - small workshops.

The well-off are, however, relatively few in number. Most people who live in Nairobi are poor and are forced to live in **shanty settlements** (photo **D**). Shanty settlements grow on land which had previously not been used. This might have been because it flooded regularly or because it was a long way from the shops and better jobs in the city centre.

Julius and Marietta come from families who live in a shanty. Most houses in a shanty settlement are small and often only have one room. They are built so close together that it can be difficult to squeeze between them. The walls are usually built from mud and the roofs from corrugated iron. Very few homes have electricity or water. Sewage runs down the tracks between houses. The tracks are muddy in the rainy season and dusty in the dry season. Uncollected rubbish lies everywhere. Shanty settlements often lack shops, schools, medical facilities and jobs.

It is almost impossible to find a permanent or a well paid job in a shanty. Some people may have stalls from which they sell food. Others find work in small workshops which they have either built by themselves or with friends. Some people may tour the local area collecting scrap and other waste materials which others will then melt down. It can then be hammered into various shapes producing objects which can be used in the home (photo **C**). These jobs are said to be **sustainable** because they involve the recycling of waste materials and they use the skills of local people.

Jua Kali workshops, recycling metal in Nairobi

C

Inside a shanty settlement – Nairobi **D**

E

Nairobi		Kip and Krista – well-off families	Julius and Marietta poorer families
Distance from city centre			
Houses	Size		
	Amenities		
Schools, shops, entertainment			
Jobs			

3 Make a larger copy of table **E**. Give it the title 'Differences between well-off and poor families living in Nairobi'. Complete the table by listing some of the differences.

Summary

Well-off people live in pleasant areas near to the city centre where there are shops, good jobs and places of entertainment. Poorer people often have to live and find work in less pleasant shanty settlements.

Rural life in the region south of Nairobi

Like many cities in the developing world, Nairobi is increasing rapidly in size. This is mainly due to

- the country's high birth rate and rapid increase in population,
- the movement of people from the surrounding rural areas who are looking for a better way of life.

As Nairobi's population increases, so too does the demand for jobs and somewhere to live. The need for jobs and houses puts more and more pressure on any land near to the city.

Before this increase in pressure on the land, much of the area to the south of Nairobi was used by animals for grazing. Herds of domestic cattle and goats belonging to the Maasai (photo **A**) grazed together with wild animals such as rhino, gazelle, zebra and, where there were more trees, giraffe.

In 1945, an area of land touching the southern edge of Nairobi became Kenya's first National Park. Today, grazing rhino and giraffe or a lion catching a gazelle can be photographed with Nairobi in the background (photo **B**). Although the Nairobi Park protects animals and brings in money, the land there cannot be used either for farming or for settlement.

To the south of the Nairobi National Park are the grazing lands of the Maasai. For centuries the Maasai have followed a **semi-nomadic** way of life. They are **pastoralists** who have to move during the long dry season in order to find grass and water for their cattle and goats. The Maasai rely upon these animals for their daily food supply.

A Maasai herding their animals

Nairobi National Park

B

C Traditional Maasai village

Until recently, most Maasai lived in small villages which consisted of between 20 to 50 huts. The huts were arranged in a circle and built from local materials. The frame consisted of wooden poles. Cow dung was mixed with river mud for the walls, and with grass for the roof (photo **C**). Each hut had a narrow tunnel like entrance. There were no windows or chimneys – just a small hole in one wall. The inside, which was partitioned into small 'rooms', was dark and full of unhealthy smoke from the fire. Cowskins were laid on the floor to act as bedding.

A thick thorn hedge surrounded each village. Its purpose was to keep out wild animals such as lions, leopards and hyenas. An opening in the hedge allowed people to pass through and was large enough for the Maasai to drive their animals through at night.

In recent times, the Maasai way of life has been changing (diagram **D**). Some of these changes have been forced upon them; others have been made through choice. The biggest changes have resulted from the loss of grazing land. As the Maasai become less free to move around with their cattle, many have made their houses more permanent. Others have decided to move away from the countryside and to live in small towns. Although many of the older Maasai still keep their traditional values and way of life, the younger generation appear to welcome the changes.

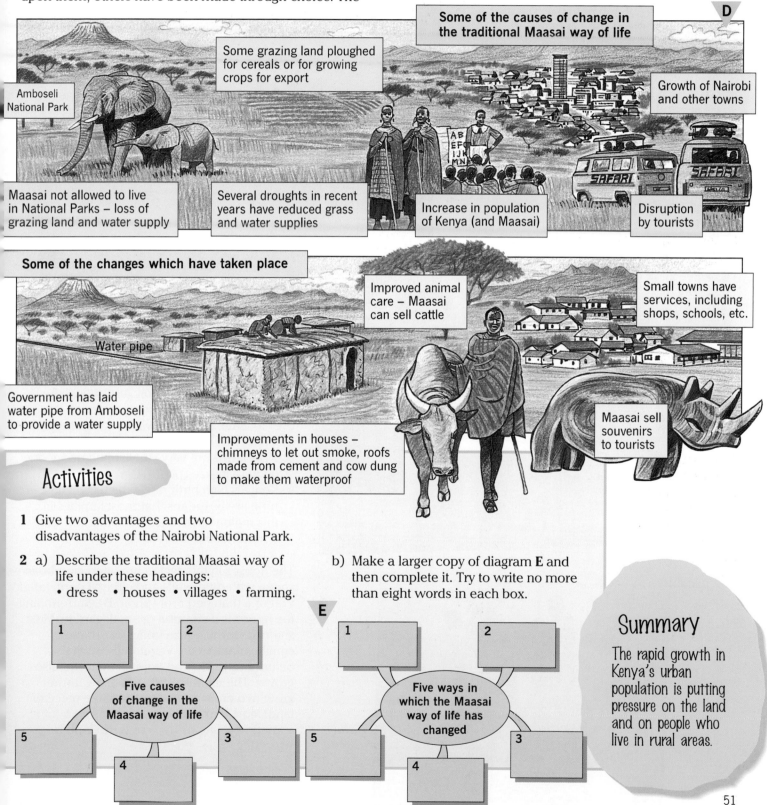

D

Some of the causes of change in the traditional Maasai way of life

Amboseli National Park

Some grazing land ploughed for cereals or for growing crops for export

Growth of Nairobi and other towns

Maasai not allowed to live in National Parks – loss of grazing land and water supply

Several droughts in recent years have reduced grass and water supplies

Increase in population of Kenya (and Maasai)

Disruption by tourists

Some of the changes which have taken place

Improved animal care – Maasai can sell cattle

Small towns have services, including shops, schools, etc.

Water pipe

Government has laid water pipe from Amboseli to provide a water supply

Improvements in houses – chimneys to let out smoke, roofs made from cement and cow dung to make them waterproof

Maasai sell souvenirs to tourists

Activities

1 Give two advantages and two disadvantages of the Nairobi National Park.

2 a) Describe the traditional Maasai way of life under these headings:
   • dress • houses • villages • farming.

b) Make a larger copy of diagram **E** and then complete it. Try to write no more than eight words in each box.

E

1
2
5
4
3

Five causes of change in the Maasai way of life

1
2
5
4
3

Five ways in which the Maasai way of life has changed

Summary

The rapid growth in Kenya's urban population is putting pressure on the land and on people who live in rural areas.

51

Subsistence farming to the west of the Rift Valley

The land to the west of the Rift Valley is highland. In many places it rises to over 2500 metres. The scenery consists of gentle ridges with a natural covering of forest. The area is ideal for both farming and settlement.

- The land, being both high and lying on the Equator, receives heavy afternoon rainfall nearly every day of the year.
- Because it is highland, the temperature is cooler and the climate healthier than in other parts of Kenya.
- Since it is on the Equator, temperatures are warm enough for crops to grow throughout the year.
- The soils, originally formed by volcanoes, are deep. They have been made even more fertile by the addition of **humus** from the natural forest.

A Shambas on a hillside in Kericho

B House and shamba

Most people who live in this highland region of Kenya are **subsistence** farmers. Subsistence farming is when farmers and their families produce only just enough for their own needs. There is rarely much left over for sale, despite having to work hard for most of each day. This is because

- Most farms are very small, often under one hectare in size (photo **A**).
- Kenya's high birth rate, which is highest in rural areas, means that most families in this region have at least six children all of whom need to be fed.

The houses in this region are often circular in shape with mud walls and either a corrugated iron or a thatched roof (photo **B**). Surrounding the house is a **shamba** or small garden. Most shambas grow maize under the shade of small banana trees. Vegetables, such as beans and yams, and tropical fruits are also grown. The climate and soils allow farmers to grow two crops a year on the same piece of land. Most farmers keep chickens and, sometimes, one or two cows. No land is wasted.

C Overcrowded local transport

Any surplus food is sold in local open air markets (photo **D**). These markets are open most mornings and are very noisy and colourful events. From early morning onwards, the sides of the rough tracks leading to the markets are packed with people. The majority have to walk, sometimes for several kilometres. Others will take advantage of any available transport (photo **C**). By mid-day, however, the heat is sufficient for people to retreat indoors and everything closes down (photo **E**). The end of the afternoon will be spent working in the shamba or, for many women, collecting fuelwood to be used in cooking stoves.

D

E

Activities

1 Draw a star diagram to show why this region is ideal for farming. Use these headings:
- height of land;
- rainfall;
- temperatures;
- soils;
- shape of land.

2 a) What is subsistence farming?
   b) Give two reasons why subsistence farming takes place in this region.

3 Diagram **F** is sketch based on photo **B**. Make a copy of the sketch and complete it by adding the following labels:

- thatch roof • circular house
- mud walls • banana trees • maize
- vegetables • shamba

4 Write a diary entry to describe the life in the day of a subsistence farming family. Use the information on these pages and diagram **G** to help you.

F

G

MONDAY 15 MAY	Male members	Female members
Early morning		
Mid-morning		
Midday		
Early afternoon		
Late afternoon		

Summary

The region west of the Rift Valley is ideal for farming. Subsistence farming is where most of the produce is needed by the farmer's family. There is very little left over for sale.

53

Commercial farming to the west of the Rift Valley

The cooler, damper, healthier climate of the region around Kericho also attracted European settlers (map **A** page 42). When the first settlers arrived they found much of the land still covered in forest. These settlers began clearing large areas of trees and replacing them with tea bushes. Today the area consists of ridge after ridge of bright, green bushes (photo **A**). Tea is grown as a **cash crop** on a **commercial** scale. Commercial farming is when crops are grown, or animals are reared, for a profit. A cash crop is one that is usually grown for export. Tea is now Kenya's most valuable export.

Brooke Bond have several huge tea estates. The largest, at Kericho (map **E**), produces 12 per cent of Kenya's tea. Young tea plants are grown in the shade for five months and are only slowly exposed to the sun and rain. The bushes are then planted in long rows. The leaves of mature bushes are still plucked (picked) by hand (photo **C**). This means that the cultivation and collection of tea takes a lot of labour (see fact file). Each plucker is given his or her own section. They start at 0700 hours and work, with two breaks, until 16.30 (or longer if they wish). The text with photos **B** and **C** describes how the leaves are plucked, and how the workers are paid and live. Diagram **D** explains what happens in the tea processing factory.

A tea estate **A**

The pluckers put the leaves into bags and are paid according to the weight of the leaves picked. The plucker returns to each bush every 17 or 18 days. This keeps the height of the bush to one metre which is ideal for later plucking. Plucking goes on throughout the year.

B

The pluckers work a nine hour day for the equivalent, in Britain, of 45p per day. They and their families are given free accommodation, education and medical care. Each tea estate has its own schools and hospital as well as shops and other amenities (map **E**). The houses have electricity, water and sewerage.

C

D In the processing factory

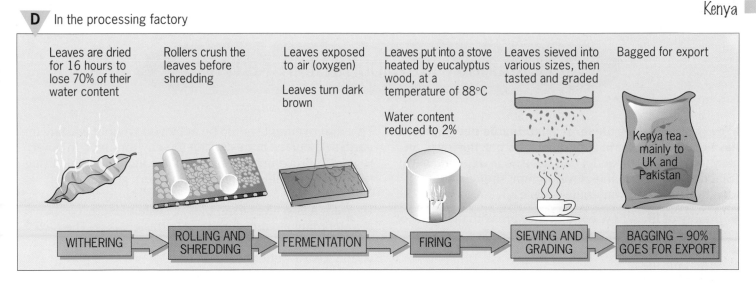

Leaves are dried for 16 hours to lose 70% of their water content

Rollers crush the leaves before shredding

Leaves exposed to air (oxygen)

Leaves turn dark brown

Leaves put into a stove heated by eucalyptus wood, at a temperature of 88°C

Water content reduced to 2%

Leaves sieved into various sizes, then tasted and graded

Bagged for export

Kenya tea - mainly to UK and Pakistan

WITHERING → ROLLING AND SHREDDING → FERMENTATION → FIRING → SIEVING AND GRADING → BAGGING – 90% GOES FOR EXPORT

Apart from tea, eucalyptus trees are also grown on the estate. These trees mature in seven years. They are used as fuel in the drying rooms of the tea-processing factory and in staff houses.

Despite the clearance of the natural forest, there is hardly any soil erosion. This is because the tea plants and the eucalyptus trees protect the ground from the heavy afternoon rains while their roots hold the soil together.

Activities

Fact file

Total area	20,000 hectares
Bushes per hectare	18,000
Number of workers' houses	9,900
Number of workers	16,000
Number of workers and families	100,000

E

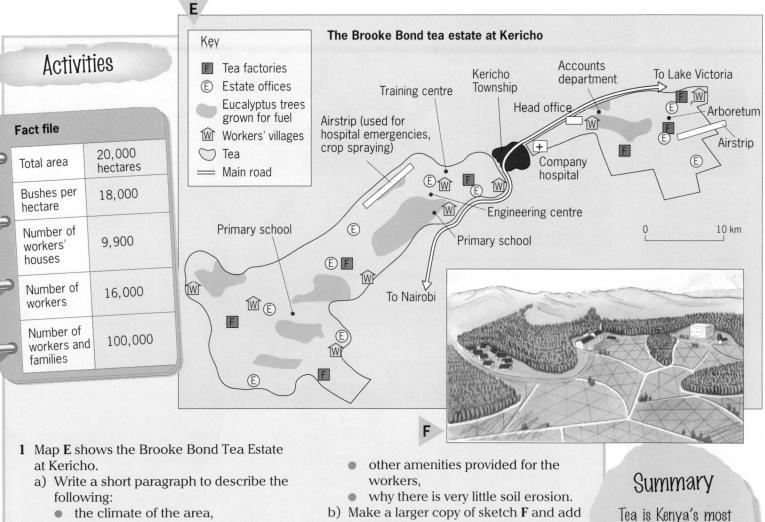

The Brooke Bond tea estate at Kericho

Key
- **F** Tea factories
- **E** Estate offices
- Eucalyptus trees grown for fuel
- **W** Workers' villages
- Tea
- = Main road

Training centre
Kericho Township
Accounts department
To Lake Victoria
Head office
Airstrip (used for hospital emergencies, crop spraying)
Company hospital
Arboretum
Airstrip
Primary school
Engineering centre
Primary school
To Nairobi
0   10 km

F

1 Map **E** shows the Brooke Bond Tea Estate at Kericho.
   a) Write a short paragraph to describe the following:
      - the climate of the area,
      - jobs done before the tea is planted,
      - how tea is collected,
      - jobs in the factory,
      - the workers' village,
      - other amenities provided for the workers,
      - why there is very little soil erosion.
   b) Make a larger copy of sketch **F** and add these labels:

   - highland • eucalyptus trees • tea bushes
   - workers village • factory • estate office

Summary

Tea is Kenya's most valuable export and is grown commercially on large estates.

Sustainable development in Kenya

In the last few years, the term '**sustainable development**' has become popular in geography. It is not, however, an easy term to explain and it can be used in different ways (diagram **A**). Sustainable development should lead, ideally, to an improvement in people's:

● **quality of life** – how content they are with their way of life and the environment in which they live;
● **standard of living** – how well off they are economically.

This improvement should be achieved without wasting the earth's resources or destroying the environment. In other words, improvements should not just benefit people living today, they should be shared with future generations.

Diagram **A** also gives examples of sustainable development already described in this chapter. The next section looks in more detail at one of these examples within Kenya – energy.

A

Sustainable development means...

. . . using natural resources without spoiling the environment

. . . developing local skills, and passing these skills on to future generations, e.g. training village people to be vets, or townspeople to make stoves

. . . developing materials/utensils which will use fewer resources, e.g. improved cooking stoves need less fuelwood.

. . . thinking about resources and the three '**re**'s:
● **re**newable resources such as wind and solar power
● **re**cycling resources as in Nairobi workshops
● **re**placing resources such as replanting eucalyptus trees on Brooke Bond's tea plantations

. . . using materials which will last for much longer, e.g. Maasai permanent houses where cement is added to cow dung to make roofs.

. . . protecting the environment, both landscapes and wildlife, so that they can be enjoyed and be useful to other generations, e.g. protecting coral reefs by creating marine reserves, and protecting rhinos and elephants in National Parks.

. . . developing technology appropriate to the skills, wealth and needs of local people

. . . allowing economic development to happen at a pace which a country can afford – giving a country too much finance means it can fall into debt

Cooking stoves

In rural Kenya, fuelwood is used by most families for cooking. Charcoal is used as a fuel in shanty settlements and in Nairobi's workshops. As Kenya's population increases so too does the demand for wood. More and more trees are cut down and not replaced. This means that

- people in rural areas have further to walk each day to find enough wood upon which to cook;
- it costs people living in towns more to buy charcoal;
- soil is eroded as there are fewer trees left to protect it from the wind and rain.

One way to save fuelwood is to design cooking stoves which are more efficient. This has been done, mainly by women, in several parts of Kenya. As traditional potters, women know that the best clay is found in river banks. Earth from termite mounds, sand, ash and water are added to the clay and the mixture is poured into a mould. It is left to dry for several days before the stove is baked in a kiln (photo **B**). By enclosing the fire, most of the heat is directed to the stove (photo **C**).

B Women unloading stoves from a kiln after firing

The Upesi stove **C**

D

Less time is spent collecting wood and in cooking, leaving more time for other activities.

Less fuel is used. Fewer trees need to be cut down in rural areas. This saves money for people needing charcoal in urban areas.

Money can be made from selling the stoves.

Less smoke is given off, reducing air pollution and improving health.

Improved cooking stoves are, therefore, an example of sustainable development.

- In their **making** they use local, renewable resources and local skills, they do not spoil the environment and cost very little.
- In their **use** they save fuel, improve people's health and save time.

Activities

1 a) What do you understand by the term 'sustainable development'?
b) Why is it important for development to be sustainable?

2 With the help of diagram **D**, explain how
a) the making, and
b) the use of cooking stoves in Kenya is a good example of sustainable development.

E

| Uses local natural resources | → | Uses and develops local skills | → | Adapted to suit the needs and wealth of local people | → | Helps to protect the environment | → | Improves people's health |

Summary

Sustainable development should lead to an improvement in people's quality of life and standard of living without wasting natural resources or spoiling the environment.

How interdependent is Kenya?

No country has everything that its people want or need. To provide these things it has to exchange goods and materials with other countries. The exchange of goods and materials is called **trade**.

A country will buy (**import**) things which are in short supply or which can be made more cheaply elsewhere. These items may be either

- **primary goods**, such as foodstuffs and minerals which are often low in value, or
- **manufactured goods**, such as machinery, which are high in value.

In order to pay for these goods, a country has to sell (**export**) materials and goods of which it has a surplus. Ideally, a country hopes to have a **trade surplus**. This means it earns more money from its exports than it has to pay for its imports. A **trade deficit** is where imports are greater than exports.

Kenya's trade is typical of most developing countries (diagram **A**). Most of its exports are low value primary goods such as tea and coffee.

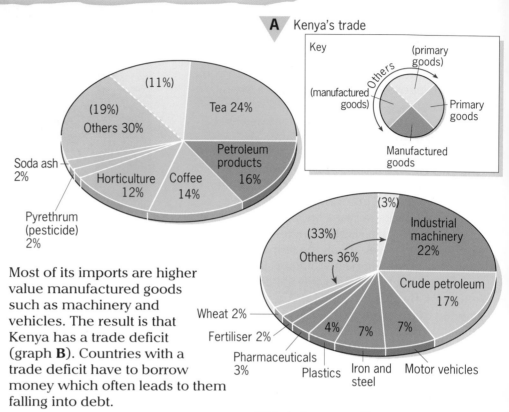

A Kenya's trade

Key

(11%)

Tea 24%

(19%)
Others 30%

Soda ash
2%

Horticulture
12%

Coffee
14%

Petroleum
products
16%

Pyrethrum
(pesticide)
2%

(3%)

(33%)
Others 36%

Industrial
machinery
22%

Crude petroleum
17%

Wheat 2%

Fertiliser 2%

Pharmaceuticals
3%

4%

7%

7%

Plastics

Iron and
steel

Motor vehicles

Most of its imports are higher value manufactured goods such as machinery and vehicles. The result is that Kenya has a trade deficit (graph **B**). Countries with a trade deficit have to borrow money which often leads to them falling into debt.

Most of Kenya's trade passes through the port of Mombasa (photo **C**). The exception are some exports, like fresh flowers and vegetables, which go through Nairobi airport. Kenya's most important **trading partner** is the United Kingdom (diagram **D**).

This is because Kenya was once a British **colony**. Colonies used to provide raw materials for colonial powers. The colonial powers then either used the raw materials or turned them into manufactured goods. These goods were then sold, at a profit, back to the colony.

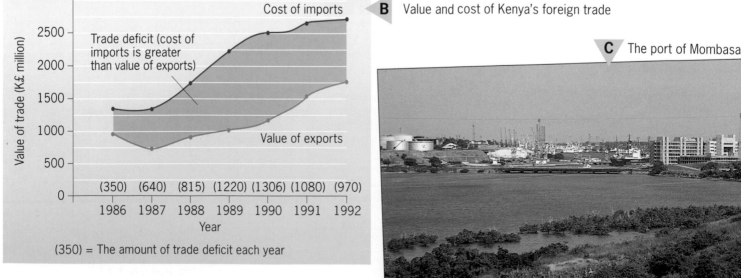

B Value and cost of Kenya's foreign trade

Cost of imports

Trade deficit (cost of imports is greater than value of exports)

Value of exports

Value of trade (K£ million)

2500

2000

1500

1000

500

0

(350) (640) (815) (1220) (1306) (1080) (970)

1986 1987 1988 1989 1990 1991 1992

Year

(350) = The amount of trade deficit each year

C The port of Mombasa

Although Kenya has been independent since 1963, most of its trade is still with economically more developed countries and the oil producing Gulf States (diagram **D**). To try to develop trade with other developing countries, Kenya has joined the **Preferential Trade Area for Eastern and Southern African States** (**PTA**). Although Kenya has a trade surplus within PTA, the amount of trade between members is small.

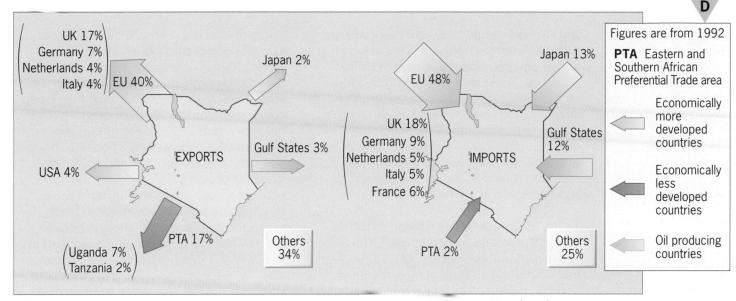

D

EXPORTS

UK 17%
Germany 7%
Netherlands 4%
Italy 4%
EU 40%

Japan 2%

Gulf States 3%

USA 4%

PTA 17%
Uganda 7%
Tanzania 2%

Others 34%

IMPORTS

EU 48%

Japan 13%

UK 18%
Germany 9%
Netherlands 5%
Italy 5%
France 6%

Gulf States 12%

PTA 2%

Others 25%

Figures are from 1992

PTA Eastern and Southern African Preferential Trade area

Economically more developed countries

Economically less developed countries

Oil producing countries

Kenya has other international links.
- It was the first developing country to become the headquarters of several United Nations agencies.
- It receives financial aid from overseas countries (mainly Germany, Japan and the United Kingdom) and international agencies (World Bank).
- Several large multinational companies, such as Brooke Bond, have invested money in developing the country's resources.
- It has also developed strong sporting and cultural links.

Activities

E

Trade	Things which are made from primary goods
Imports	Buying goods or materials from another country
Exports	Raw materials such as foodstuffs and energy resources
Trade surplus	Cost of imports is greater than money obtained from exports
Trade deficit	The exchange of goods and materials between counties
Primary goods	Money from exports is more than money paid for imports
Manufactured goods	Selling goods made in or obtained from a country

1 Complete the definitions in table **E** by matching the beginnings on the left with the correct ending from the list on the right.

2 Copy and complete table **F** which describes the major features of Kenya's trade. Do not use the 'Others' section from diagram **A**.

F

Trade (1992)	Exports (in order)	Imports (in order)
Primary goods	1 ___ 2 ___ 3 ___	1 ___
Manufactured goods	1 ___	1 ___ 2 ___ 3 ___
Value	£Kmn	£Kmn
Main countries	1 ___ 2 ___ 3 ___	1 ___ 2 ___ 3 ___
% with developed countries	%	%
% with developing countries	%	%

Summary

Countries need to be interdependent if they are to share in the world's resources. Countries like Kenya which export primary goods do not develop as quickly economically as those which export manufactured goods.

How developed is Kenya?

By now you should be aware that there are many differences between living in Kenya and in other countries such as the United Kingdom. These differences may include ethnic groups, dress, housing, religion, language, jobs and wealth. You should also be aware that countries are at different levels of development. The term development is not, however, easy to define. The most common meaning, though not always the most accurate, is based upon wealth.

To make comparisons between countries easier, the wealth of a country is given by its **gross national product** (**GNP**). Remember that GNP is the amount of money earned by a country divided by its total population.

Diagram **A** is based on GNP. It shows that Kenya is one of the poorest countries in the world although it is about average for Africa. Based on wealth, therefore, Kenya is considered to be one of the least developed countries in the world. What diagram **A** does not show, however, are the differences in wealth within Kenya itself. There are many Kenyans who are wealthy and have a high standard of living. Unfortunately the majority of people, especially those living in urban shanty settlements or isolated rural areas, are very poor (pages 48–49). The gap between the rich and the poor is a major problem in Kenya as it is in other developing countries.

Wealth is not the only way to measure, or describe, development. Development can be measured by differences in population, health, education, jobs and trade (diagram **B**). Yet each of these measures can be linked to wealth. Without wealth a country is unlikely to have enough money to spend on such things as schools, health care and birth control.

When applying most of these measures to a developing country such as Kenya it is easy to build up a negative and often incorrect picture of that place. More important than wealth should be the **quality of life**. Many Kenyans, especially those living in shanty settlements in Nairobi, are very poor. Yet the majority appear to be cheerful, relaxed and are nearly always prepared to help visitors or their neighbours (photo **C**). Socially they appear more developed and less selfish than many people in so-called 'richer ' countries. Quality of life is hard to measure, in comparison to wealth, birth rates and trade.

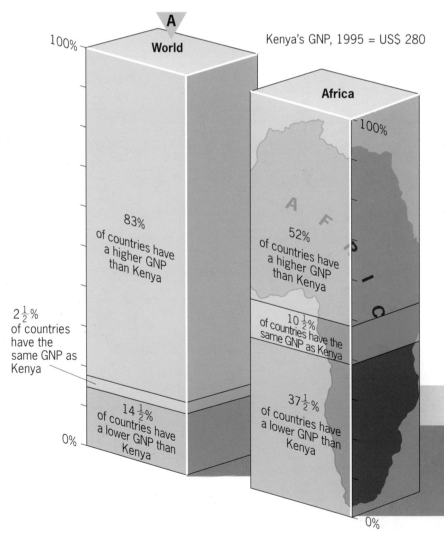

A

Kenya's GNP, 1995 = US$ 280

World
100%

83% of countries have a higher GNP than Kenya

$2\frac{1}{2}\%$ of countries have the same GNP as Kenya

$14\frac{1}{2}\%$ of countries have a lower GNP than Kenya

0%

Africa
100%

52% of countries have a higher GNP than Kenya

$10\frac{1}{2}\%$ of countries have the same GNP as Kenya

$37\frac{1}{2}\%$ of countries have a lower GNP than Kenya

0%

B

Jobs	Trade	Population	Health	Education
Most Kenyans are farmers, or are involved in primary activities. Relatively few are employed in manufacturing or services (apart from tourism).	Kenya's main exports are primary goods (raw materials) which it sells at a low price. Its main imports are manufactured goods which it has to buy at higher prices.	Kenya has one of the highest birth rates in the world. It also has more young children dying (high infant mortality), adults dying at a younger age (shorter life expectancy) and a faster population increase than a developed country.	Kenya has relatively little money to spend on training doctors and nurses, and in providing hospitals and medicine.	Half of Kenya's population are of school age. Despite a lot of money spent on education, fewer people can read than in a developed country (low literacy rate). Girls do not receive the same education as boys.

One of the first photos in this unit (photo **B**, page 38) showed victorious Kenyan athletes. The unit ends by reminding you that when it comes to long distance races, as just one example, Kenyans are the world's best and the **most** developed.

C

Activities

1 a) Name the seven measures shown in diagram **D** which can be used to describe a country's level of development.
   b) Explain how each one can be used to measure development.
   c) Which ones of the seven suggest that Kenya is still a developing country?
   d) Why is GNP not always a good measure of development?

D

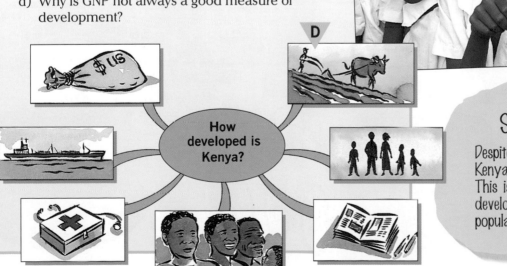

How developed is Kenya?

Summary

Despite its social and cultural development, Kenya is seen as a developing country. This is due to its limited economic development, lack of wealth and rapid population growth.

61

4 Italy

What is Italy like?

Most of us have a mental picture or image of what a country is like. A group of geography pupils were asked what pictures came to mind when they thought about Italy. The most popular ones that they came up with are shown in drawing **A** below. Look carefully at the drawing. How many of the pictures can you put names to? What other ones would you have thought of?

Italy, in fact, is one of Europe's best-known countries. It is often in the news and an increasing number of people have visited the country on holiday. The Italian passion for fashion, fast cars and football is well known throughout the world.

Look at drawing **B** which compares Italy with the United Kingdom. Notice that in many ways Italy is very similar to the UK. If you look closely however, you will see that there are also some quite important differences.

It is easy to find us on a map. We are situated in southern Europe. Our country is long and narrow and is almost surrounded by the Mediterranean Sea. The islands of Sicily and Sardinia are also part of Italy.

A

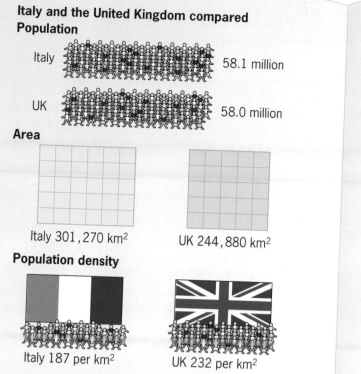

B

Italy and the United Kingdom compared
Population

Italy   58.1 million

UK   58.0 million

Area

Italy 301,270 km²

UK 244,880 km²

Population density

Italy 187 per km²

UK 232 per km²

	Italy	UK
In secondary education	74%	83%
Cars (per 100 people)	43	35
Televisions (per 100 people)	28	37
Tractors (in millions)	1.4	0.5
Tourists (in millions)	56	57
Urban population	72%	94%
Life expectancy (in years)	76	76

Activities

1 a) Captions for six of the pictures in drawing **A** are given below. Match the beginnings with the correct endings to find them.
b) Write down a caption for each of the other four pictures.
c) Work with a partner and make a list of any other things that come to mind when you think of Italy.

2 Copy four pieces of information from drawing **B**. Next to each one explain what you think it tells you about Italy.

3 a) Give four pieces of information that show Italy to be similar to the UK.
b) Give four pieces of information that show Italy to be different to the UK.

4 Draw a poster to show what Italy is like. Use a sheet of A4 paper or a double page in your exercise book. Try to include:
- some drawings
- some captions
- some factual information

Make your poster as eyecatching, interesting, and colourful as you can.

C

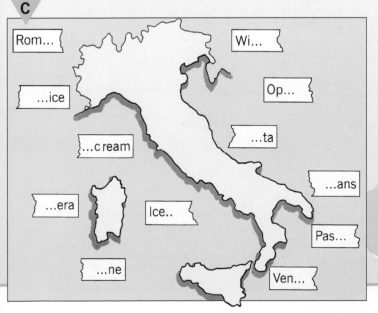

Rom... Wi... ...ice Op... ...cream ...ta ...ans ...era Ice.. Pas... ...ne Ven...

Summary   Italy is a beautiful country with a long and interesting history. Italians have developed their own customs and way of life.

What are Italy's main physical features?

Earthquake shock hits Naples area. 4000 feared dead

Villages threatened as Etna erupts again

POMPEII DESTROYED BY VESUVIUS ERUPTION (AD 79)

Earthquake blamed for Sicily landslides. 2500 left homeless for Christmas

Headlines like the ones above are quite common in Italy. As map **A** shows, there have been four major **earthquakes** in the last 25 years. There have also been several minor ones. Volcanic activity is even more frequent. Italy has Europe's only **active volcanoes** and eruptions happen every year. Problems like this are called **natural hazards**.

Earthquakes and volcanic eruptions are caused by movements of the **earth's crust**. These movements also help to build mountains. Over three quarters of Italy is hilly or mountainous.

Geologically, the country is quite young and most mountain building activity happened between 40 and 20 million years ago. This has left Italy with some spectacular scenery as the landscape has not yet been worn away and rounded off as much as it has in older places like Britain.

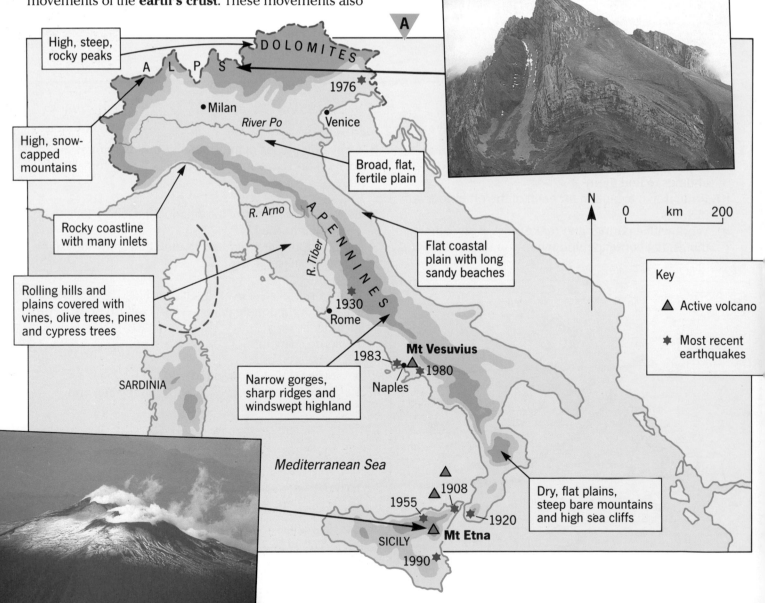

A

High, steep, rocky peaks

DOLOMITES

ALPS

1976

• Milan

River Po

Venice

High, snow-capped mountains

Broad, flat, fertile plain

Rocky coastline with many inlets

R. Arno

APENNINES

R. Tiber

Flat coastal plain with long sandy beaches

Rolling hills and plains covered with vines, olive trees, pines and cypress trees

1930
Rome

Mt Vesuvius

1983

1980

Narrow gorges, sharp ridges and windswept highland

Naples

SARDINIA

N

0 km 200

Key

△ Active volcano

★ Most recent earthquakes

Mediterranean Sea

1908

1955

1920

Dry, flat plains, steep bare mountains and high sea cliffs

Mt Etna

SICILY

1990

The earth's crust is broken into several enormous sections called **plates**. The boundaries and movement of plates near to Italy is very complicated and is not yet fully understood. Diagram **B** shows their probable movement. Scientists believe that as the plates move towards each other, heat from the friction between them makes the rock melt. The liquid rock rises upwards and erupts through holes in the earth's surface as volcanoes. Etna and Vesuvius are two such volcanoes.

The force of the two plates coming together puts tremendous pressure on the rock layers. This pressure causes them to bend and **fold**. The Apennines and Alps are **fold mountains** caused by rock being pushed upwards and folding. The movement of plates scraping together also makes the ground shake and sets off earthquakes.

B How some of Italy's physical landscape may have been formed.

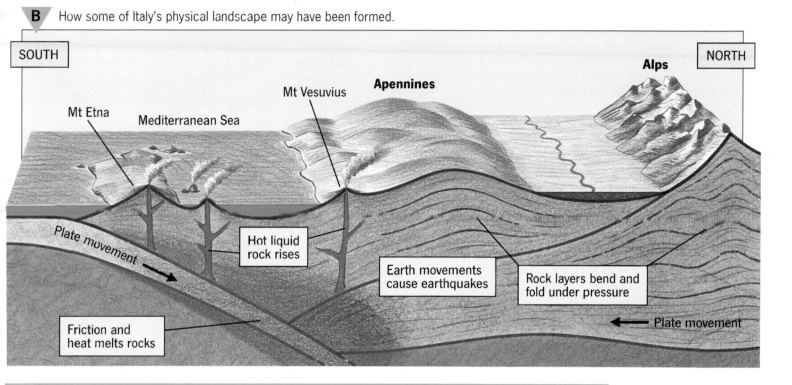

SOUTH · NORTH

Alps

Apennines

Mt Vesuvius

Mt Etna · Mediterranean Sea

Plate movement

Hot liquid rock rises

Earth movements cause earthquakes

Rock layers bend and fold under pressure

Plate movement

Friction and heat melts rocks

Activities

1 Why are earthquakes and volcanic eruptions called natural hazards?

2 Describe how the Alps were formed by sorting out the boxes below into the correct order. Link your boxes with arrows and add a title.

Pressure causes rocks to bend and fold

Two plates move toward each other

Rock layers put under pressure

Upward movement causes Alps to form

3 a) What are the dates of the four most recent earthquakes?
b) Describe how earthquakes happen.

4 a) Make a copy of cross section **C**.
b) Add labels to the boxes to show how Mount Etna erupts.
c) Give your drawing a title.

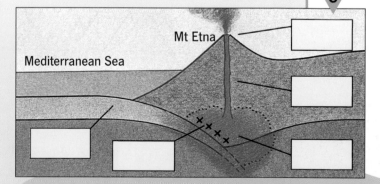

Mt Etna

Mediterranean Sea

C

Summary Italy is a mountainous country with little flat land. The landscape is mainly a result of earth movements happening in recent times.

Where do people live in Italy?

As we have seen, Italy and the United Kingdom are similar in many ways. They both have a population of about 58 million. They are a similar size in area, and in both countries, the population is unevenly spread out.

Map **A** shows where people live in Italy. It is easy to see that some areas have a lot of people and some have very few. The most crowded places seem to be toward the north and near the coast. The least crowded are the south, the centre and the extreme north.

There are good reasons for this **population distribution**. On the North Italian Plain, for example, the land is flat and ideal for farming. Building is easy and big industrial centres such as Milan and Turin have grown up. The availability of food, jobs and other facilities have encouraged people to live here so the area has become **densely populated**. On the other hand, life in many parts of the south of Italy is very difficult. The area is hot and dry and very rugged. Farming is not easy and industry has been slow to develop because of the long distances to the main markets of the north. Conditions have discouraged people from living here and the area is **sparsely populated**.

Some of the things that affect where people live are shown in drawing **B** below. Look at the photos on the opposite page. Their locations are shown on the map. For each one in turn try to work out why the area is likely to be densely populated or sparsely populated.

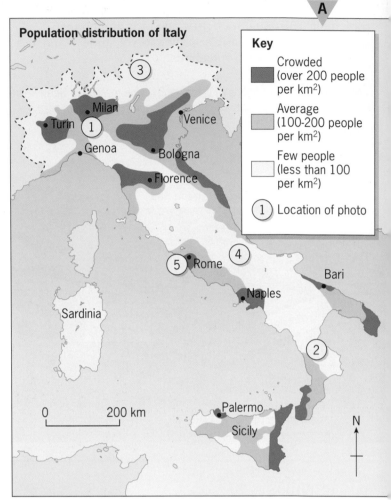

A

Population distribution of Italy

Key

Crowded (over 200 people per km²)

Average (100-200 people per km²)

Few people (less than 100 per km²)

① Location of photo

More than three quarters of the Italian people live in towns and cities. Rome, the capital and largest city, has nearly 3 million inhabitants. Like most cities in Italy it has magnificent buildings and a rich and interesting history. It is also nearly always crowded, noisy and congested.

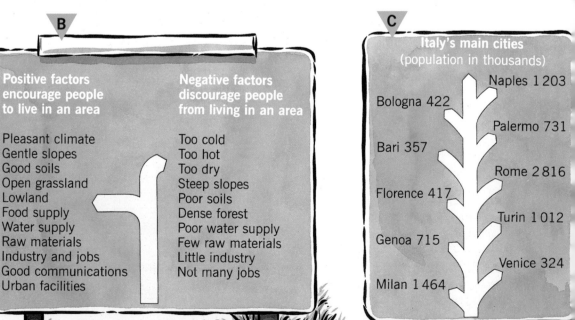

B

Positive factors encourage people to live in an area	Negative factors discourage people from living in an area
Pleasant climate	Too cold
Gentle slopes	Too hot
Good soils	Too dry
Open grassland	Steep slopes
Lowland	Poor soils
Food supply	Dense forest
Water supply	Poor water supply
Raw materials	Few raw materials
Industry and jobs	Little industry
Good communications	Not many jobs
Urban facilities	

C

Italy's main cities (population in thousands)

Naples 1 203
Bologna 422
Palermo 731
Bari 357
Rome 2 816
Florence 417
Turin 1 012
Genoa 715
Venice 324
Milan 1 464

1 Milan on the North Italian Plain

2 Campania in South Italy

5 Rome

3 Alps

4 Apennines

Activities

1 a) Is the population of Italy evenly or unevenly distributed?
  b) Where are the most crowded places?
  c) Where are the areas that are least crowded?

2 For each of the five photos above:
  a) Name the area shown in the photo.
  b) Say if it is crowded or sparsely populated.
  c) Suggest reasons for your answer. Drawing **B** will help you.

3 a) List the cities from diagram **C** in order of size. Give the biggest first.
  b) Which of the cities are on the coast?
  c) Which of the cities are on the densely populated North Italian Plain?

4 a) What is the population of Italy?
  b) What proportion of Italians live in cities?
  c) What are the advantages of living in a city like Rome?
  d) What are the disadvantages of living in a city like Rome?

Summary

People are not spread evenly through Italy. Some areas are very crowded whilst others are almost empty. Most Italians live in cities.

Wish you were here?

Italy is one of Europe's top tourist destinations. Over 55 million people visit the country each year. Most of them come from Germany, France, the UK and the USA. Most Italians also holiday in their own country. The majority of them go to the seaside or mountains.

A

B

There are three main reasons for Italy's popularity with tourists.

 Firstly, Italy has a long and fascinating history. The country is full of archeological sites and historic remains. The ruins of ancient Roman buildings such as imperial palaces, temples, arches and simple villas are to be seen in many places.

 The second reason is the landscape. Italy is a beautiful country with very varied scenery. There are snow-capped mountains, thickly wooded hills and spectacular volcanoes. Along the coast there are miles of sandy beaches, steep cliffs and rocky islands.

Thirdly, and perhaps most important of all, is the weather. Almost all of Italy enjoys a **Mediterranean climate**. This gives hot summers, warm winters and long hours of sunshine throughout the year. Only in winter is there enough rain to worry about.

C Some attractions of Italy

Glaciers
Lovely islands
Peaceful countryside
Interesting cities
Magnificent architecture
Coastal cliffs
Art galleries
Museums
Football
Sport

Watersports
Wine
Good food
Ski–ing
Mountaineering
Thermal springs
Fine hotels
Holiday entertainment
Good shopping

Plenty of sunshine
Warm temperatures
Little rain
Good climate
Fine beaches
Warm seas
Fascinating history
Spectacular mountains
Volcanoes
Lakes

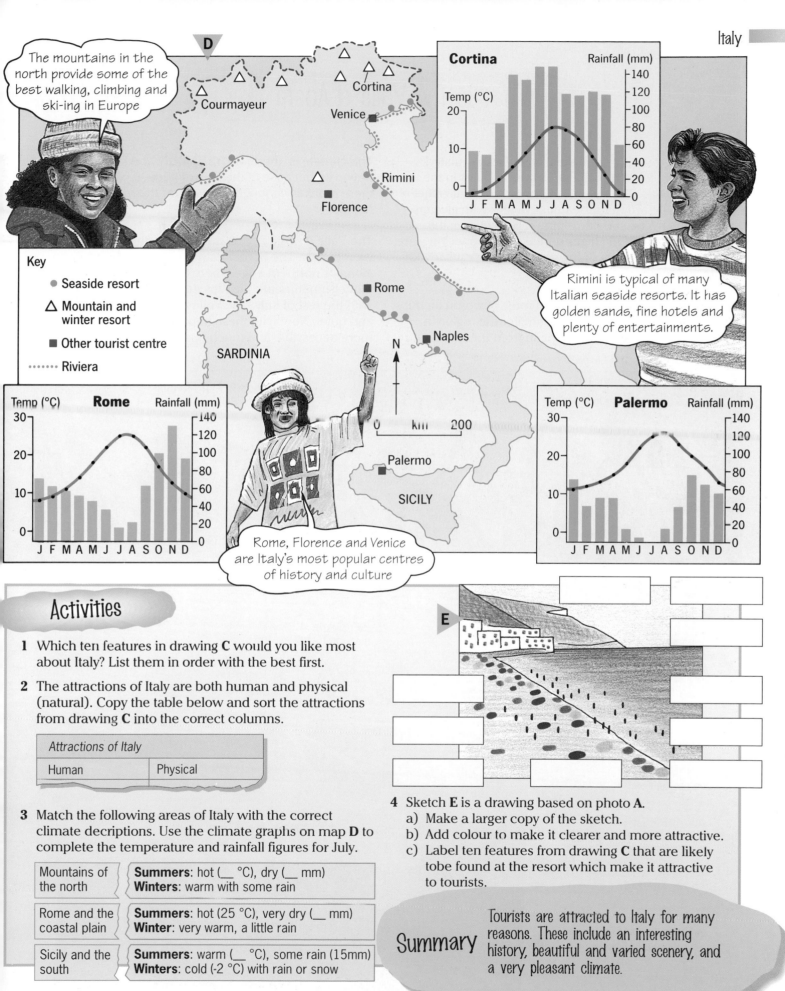

The mountains in the north provide some of the best walking, climbing and ski-ing in Europe

Cortina Rainfall (mm)

Rimini is typical of many Italian seaside resorts. It has golden sands, fine hotels and plenty of entertainments.

Rome Rainfall (mm)

Palermo Rainfall (mm)

Rome, Florence and Venice are Italy's most popular centres of history and culture

Key
- ● Seaside resort
- △ Mountain and winter resort
- ■ Other tourist centre
- ⋯⋯ Riviera

Courmayeur · Cortina · Venice · Rimini · Florence · Rome · Naples · Palermo · SARDINIA · SICILY

Activities

1 Which ten features in drawing **C** would you like most about Italy? List them in order with the best first.

2 The attractions of Italy are both human and physical (natural). Copy the table below and sort the attractions from drawing **C** into the correct columns.

Attractions of Italy	
Human	Physical

3 Match the following areas of Italy with the correct climate decriptions. Use the climate graphs on map **D** to complete the temperature and rainfall figures for July.

Mountains of the north	**Summers**: hot (__ °C), dry (__ mm) **Winters**: warm with some rain
Rome and the coastal plain	**Summers**: hot (25 °C), very dry (__ mm) **Winter**: very warm, a little rain
Sicily and the south	**Summers**: warm (__ °C), some rain (15mm) **Winters**: cold (-2 °C) with rain or snow

4 Sketch **E** is a drawing based on photo **A**.
a) Make a larger copy of the sketch.
b) Add colour to make it clearer and more attractive.
c) Label ten features from drawing **C** that are likely tobe found at the resort which make it attractive to tourists.

Summary Tourists are attracted to Italy for many reasons. These include an interesting history, beautiful and varied scenery, and a very pleasant climate.

What is the Valle d'Aosta like?

The Valle d'Aosta is the smallest and least populated Italian region. It is a mountainous area and lies in the heart of the Alps. The region is famous for its magnificent scenery. There are many beautiful valleys with glaciers, fast-flowing rivers, and fine alpine forests. Above the valleys are some of the highest peaks in Europe. They are snow-covered throughout the year and are popular with skiers and mountaineers.

The weather here is better than in most places in the Alps. The high mountains to the north shelter the region in winter from bad weather coming in from the Atlantic.

The climate is, therefore, mainly dry and sunny. In winter, however, because of the high altitude, it can be cold and there is alwasy enough snowfall for the many ski-ing resorts.

The Valle d'Aosta region is located in the extreme north-west of Italy. The border with France and Switzerland runs along its northern edge. There are now two great road tunnels underneath the Alps that link these two countries with the rest of Italy. The Valle d'Aosta was once isolated and quiet. Now with these improved communications it is more accessible and much busier.

A Valle d'Aosta region

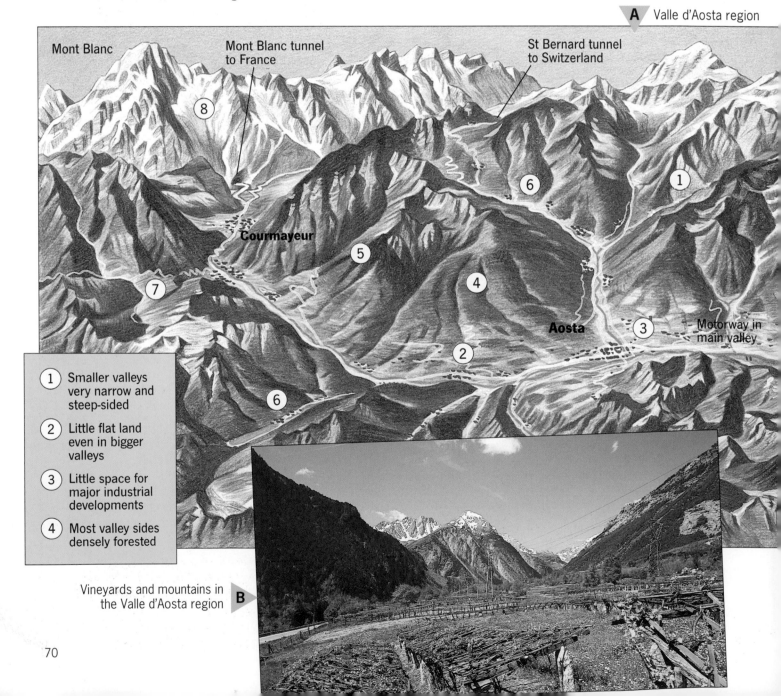

Mont Blanc

Mont Blanc tunnel to France

St Bernard tunnel to Switzerland

Courmayeur

Aosta

Motorway in main valley

① Smaller valleys very narrow and steep-sided

② Little flat land even in bigger valleys

③ Little space for major industrial developments

④ Most valley sides densely forested

Vineyards and mountains in the Valle d'Aosta region **B**

C Map of the Valle d'Aosta region

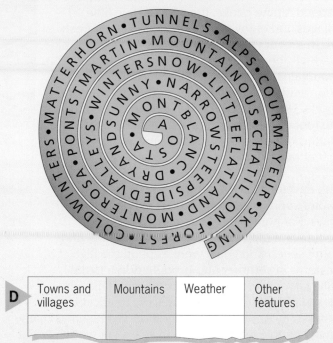

Key

- Highland
- Lowland

0 ___ 100 km

Switzerland

Mt. Blanc

The Valle d'Aosta region

•Milan

•Turin

ITALY

Genoa

France

Mediterranean Sea

N

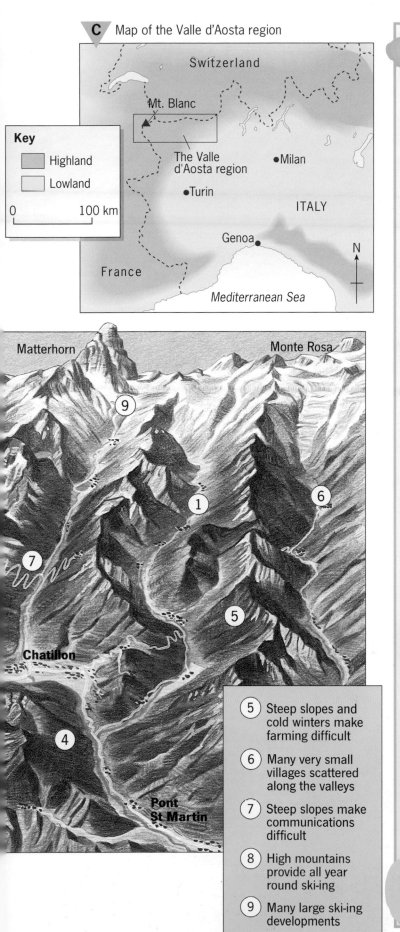

Matterhorn

Monte Rosa

Chatillon

Pont St Martin

⑤ Steep slopes and cold winters make farming difficult

⑥ Many very small villages scattered along the valleys

⑦ Steep slopes make communications difficult

⑧ High mountains provide all year round ski-ing

⑨ Many large ski-ing developments

Activities

1 Look at sketch **A** which shows some features of the Valle d'Aosta region. Find seventeen of these features in the word spiral. Sort them under the headings in a copy of table **D** below. Read the spiral from the centre outwards.

MATTERHORN • TUNNELS • ALPS • PONTSTMARTIN • MOUNTAINOUS • COURMAYEUR • WINTERSNOW • LITTLEFLATLAND • CHATILLON • SUNNY • NARROWSTEEPSIDED • SKIING • MONTBLANC • FOREST • DRYANDSUNNY • AOSTA • COLDWINTERS • MONTEROSA • DRYANDVALLEYS

D

Towns and villages	Mountains	Weather	Other features

2 What improvements in communications have helped make the region more accessible and busier?

3 Why is the area good for ski-ing? Give at least four reasons.

4 Draw a simple sketch map to show the location of the Valle d'Aosta region. Drawing **A** on page 66 and map **C** will help you. Mark on and name:
   a) France, Switzerland and Italy,
   b) Aosta and three other towns,
   c) the Alps, Apennines and Mediterranean Sea,
   d) River Po, North Italian Plain and Mont Blanc,
   e) Add a key, scale line and direction and two road tunnels.

5 What is the distance from Aosta to
   a) Milan,
   b) Mediterranean Sea.

Summary The Valle d'Aosta region is a mountainous part of north-west Italy. The area is sparsely populated and there are few towns.

Winter sports: the good and the bad

Over the years, the Valle d'Aosta region has become increasingly popular with tourists. Most visitors come to see the beautiful snow-capped Alpine peaks, the picturesque little villages and enjoy peace, quiet and clean fresh air. In recent years, however, ski-ing has gradually become a main attraction. The high mountains, good winter snowfall and generally fine weather make the area perfect for winter sports. Once quiet centres like Courmayeur at the foot of Mont Blanc, and Cervinia below the Matterhorn have become bustling little towns that attract skiers from all over the world.

A

The development of tourism, and especially winter sports, has changed the region in many ways. For hundreds of years the main occupation of the area was farming. Cattle and some sheep and goats were kept on the high pastures in summer and brought down to the valley floor in winter. Alpine communities were close-knit, self-contained and slow to accept modern technology. Out-migration increased as many young people left the region in search of work and the attractions of city life. With the building of the tunnels and new roads, however, the area became much easier to get to. As the tourist boom began, new facilities had to be built to cater for their needs.

As photo **B** shows, these developments have brought many benefits to the area. Tourism has become an important source of wealth and standards of living for most local people have improved.

Unfortunately there have also been problems. Many people are concerned that winter sports facilities, in particular, are spoiling the countryside and permanently damaging the environment. Cartoon **C** shows some of these problems.

B Some benefits of tourism

More jobs

Money comes into area

New tourist amenities which locals can use

More people to meet – more interesting life

Fewer people leave the area

Better paid jobs

Improved roads

Better quality of life

C Some problems of tourism

Ugly ski-ing facilities spoil mountainside
· Danger from falling rocks
· New building work a nuisance
· Increased chance of floods
· Hotels spoil landscape
· Traffic jams
· Forests cleared
· Wildlife frightened away
· Plant life damaged

The ski-ing's great but is it worth all the damage?

Activities

1 a) What five reasons make the Valle d'Aosta popular with tourists?
b) What three reasons help make the region good for winter sports?

2 Describe three features of life in the Valle d'Aosta before the arrival of large scale tourism.

3 Which benefits from photo **B** do you think would be most useful to a teenager about to leave school?

4 Match each of the labels from the notice board with the correct number from cartoon **C**.

5 a) Make a copy of table **D**.
b) Tick the boxes for each effect of tourism.
c) Colour the benefits in green and the problems in red.
d) Choose any two problems and suggest what could be done to reduce them.

6 Do you think that winter sports developments should be encouraged or discouraged? Give reasons for your answer. Write about half a page or work with a partner for this question and present your arguments in a debate.

D

Effect	People	Environment
Increased wealth		
People to meet		
Traffic jams		
New jobs		
Deforestation		
Flooding		
Things to do		
Loss of wildlife		
Reduced out-migration		

Summary The Valle d'Aosta region is very popular with tourists particularly in winter. Tourism has brought many benefits but it has also caused problems.

Italy's Industrial Triangle

A

olivetti

ALFA ROMEO

ZANUSSI

FIAT

PIRELLI

UNITED COLORS OF BENETTON.

You probably recognise the names on the left. They belong to well-known Italian companies that sell their products all around the world. These companies have their factories and main offices in an area that is called the Industrial Triangle.

The Industrial Triangle is located just south of the Valle d'Aosta region. It lies at the western end of Italy's largest area of lowland, the North Italian Plain. The cities of Milan, Turin and Genoa are at the three corners of the 'triangle'. Milan is the commercial and business centre of the region. Turin is an industrial city and the home of Fiat, one of the world's largest companies. Genoa is a major port on the Mediterranean Sea and an outlet for exports to foreign countries.

Industry developed in this area much later than in most other countries in Western Europe. It was not until the late 1900s that industrial development really took off in the region. This was largely because Italy has few natural resources. A lack of coal, in particular, caused problems until other forms of energy were developed. Now, energy for the region comes mainly from hydro-electric power stations, natural gas and a small number of nuclear power stations. Even now, some electricity has to be imported from France because these power sources are not sufficient to meet Italy's growing needs.

In the last 50 years industrial growth has been very rapid. Italy is now one of the seven richest nations in the world and the Industrial Triangle has become the wealthiest region in the whole country.

Look carefully at map **B**. It shows the main features of Italy's Industrial Triangle. See how many reasons you can find for the area being so attractive to industry.

Fashion / Shoes / Cars / Machinery / Textiles / **Industry in Italy** / Steel / Food / Ceramics / Chemicals

Activities

1 Draw a simple sketch map of the region shown in map **B**. Follow these instructions.
  a) Draw in the coastline and name the Mediterranean Sea.
  b) Mark the areas of highland and name the Alps and Apennines.
  c) Mark with a red dot and name Milan, Turin and Genoa.
  d) Draw in and name the Industrial Triangle.
  e) Shade the highland brown, the lowland green and the sea blue.
  f) Add a title and key to your map.

2 Give three facts about the location of the Industrial Triangle.

3 a) Why was Italy slower to industrialise than other countries in Western Europe?
  b) What main sources of energy are now used in the Industrial Triangle?

4 Make a larger copy of diagram **C** and complete it using information from map **B**.

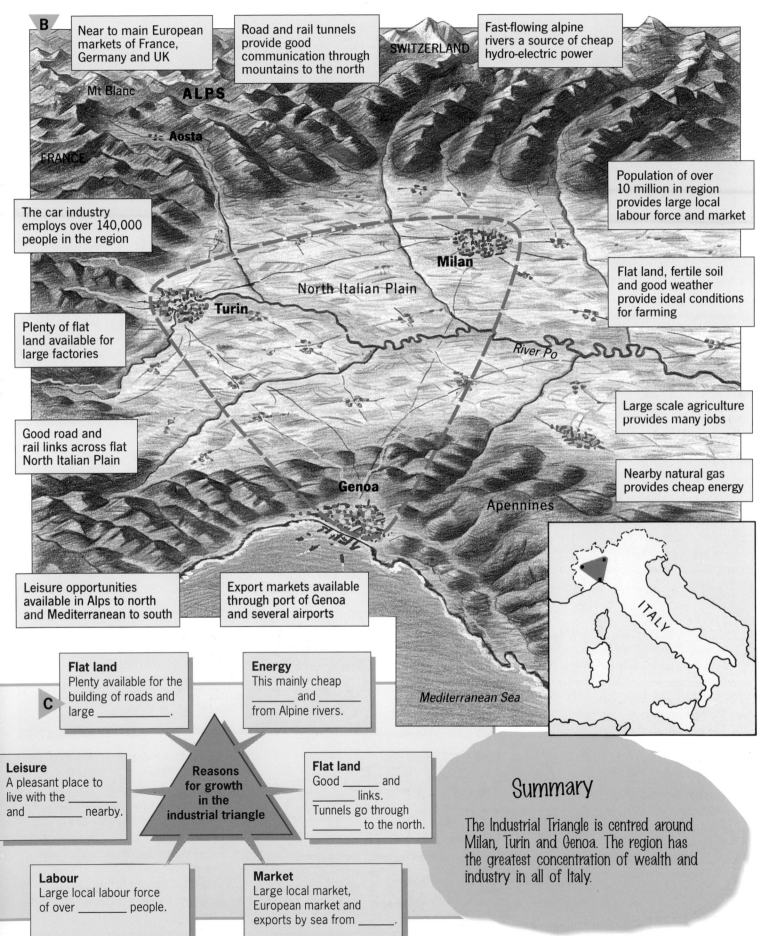

B

Near to main European markets of France, Germany and UK

Road and rail tunnels provide good communication through mountains to the north

SWITZERLAND

Fast-flowing alpine rivers a source of cheap hydro-electric power

Mt Blanc

ALPS

Aosta

FRANCE

The car industry employs over 140,000 people in the region

Population of over 10 million in region provides large local labour force and market

Milan

North Italian Plain

Turin

Flat land, fertile soil and good weather provide ideal conditions for farming

Plenty of flat land available for large factories

River Po

Good road and rail links across flat North Italian Plain

Large scale agriculture provides many jobs

Genoa

Nearby natural gas provides cheap energy

Apennines

Leisure opportunities available in Alps to north and Mediterranean to south

Export markets available through port of Genoa and several airports

ITALY

Mediterranean Sea

C

Flat land
Plenty available for the building of roads and large _____.

Energy
This mainly cheap _____ and _____ from Alpine rivers.

Leisure
A pleasant place to live with the _____ and _____ nearby.

Reasons for growth in the industrial triangle

Flat land
Good _____ and _____ links.
Tunnels go through _____ to the north.

Summary

The Industrial Triangle is centred around Milan, Turin and Genoa. The region has the greatest concentration of wealth and industry in all of Italy.

Labour
Large local labour force of over _____ people.

Market
Large local market, European market and exports by sea from _____.

Milan: a good place to live?

Buon giorno,

my name is Gennaro and I live with my family on the northern outskirts of Milan. We all like living in Milan. It's very big, very busy and there is always plenty to do.

Where we live is quite industrial but the city centre around the Piazza del Duomo is beautiful with many fine buildings. The cathedral is really spectacular. It has a hundred spires, a thousand statues and can hold 40 000 people. Next to it is the Galleria Vittoria Emmanuelle which is a glass-roofed arcade with lovely shops and restaurants. My parents like to go to the nearby La Scala theatre, one of the world's great opera centres.

My favourite place in Milan is the San Siro Stadium. It is the home of two of the world's top football teams. The World Cup final was held there in 1994 but unfortunately Italy lost! My sister came with me to the final but she really prefers night-clubs. There are plenty of them in Milan.

Milan is Italy's most important industrial and financial centre so I should have a good choice of jobs when I leave school. There are big engineering, chemical and food processing factories not far from where we live. I think I would like to work for Alfa Romeo making cars. My sister is trying to get into the fashion industry. Milan is famous for fashion clothing. Benetton, Armani and Gucci all have their headquarters in the city.

The roads around here are very good. Within a couple of hours we can be into the Alps for ski-ing or down to the Mediterranean to enjoy the beaches. That makes Milan an even better place to live.

A Milan cathedral and city centre

I ♥ Milan

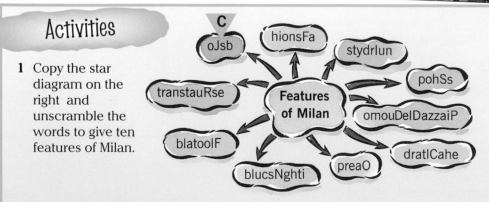

B San Siro Stadium

Activities

C

oJsb · hionsFa · stydrlun · pohSs · omouDelDazzaiP · dratlCahe · preaO · blucsNghti · blatoolF · transtauRse

Features of Milan

1 Copy the star diagram on the right and unscramble the words to give ten features of Milan.

2 Name five industries that may be found in Milan.

3 What places might easily be visited from Milan at weekends?

4 Read carefully what Francesca has to say about living in Milan. Copy the four headings from diagram **E** and add notes to each one.

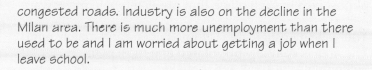

Buon giorno,

my name is Francesca. I live with my parents in Gorgonzola just 18 kilometeres west of Milan's city centre.

Gorgonzola is famous for its cheese. When my parents moved here it was a quiet and pretty little town with vineyards and lovely countryside around it. Now we are just about part of Milan and the place is crowded and noisy.

The traffic around here is bad enough but in Milan centre it is even worse. At peak times they have traffic grid lock and it is almost impossible to cross the city. The cars also combine with industry to cause air pollution. Conditions are worst in winter. The cold, still air from the mountains traps the exhaust fumes and factory smoke. It produces a choking smog which often sits over the city for days.

Like every other big city, Milan has its problems. We think that many of our problems are because the city has grown too quickly. There doesn't seem to have been enough time to plan carefully and we have lost much of what was good and attractive about Milan. A lot of open space and parkland has gone in the last few years. This has mostly been replaced by huge blocks of flats, ugly estates and even more congested roads. Industry is also on the decline in the Milan area. There is much more unemployment than there used to be and I am worried about getting a job when I leave school.

We are thinking of moving away from Milan soon. My parents are looking for work and a place to live in the lovely Valle d'Aosta region just north-west of here. We think that is a better place to live than Milan.

D Traffic congestion and pollution in Milan city centre

Traffic

Air pollution

Planning

Problems in Milan

Jobs and industry

E

5 If you lived in Milan would you
a) prefer to stay there or
b) prefer to move away to somewhere like the Valle d'Aosta region?
Give reasons for your answer.

Summary

Milan is one of Italy's largest and most interesting cities. In recent years, rapid growth has caused problems for the people living there.

Two regions compared

Both the Valle d'Aosta region and the Industrial Triangle are located in north-west Italy. Whilst being located almost next to each other, they are different in many ways.

The Valle d'Aosta region is mountainous and mainly rural. The area is **sparsely populated** and most of the settlements are small and crowded along narrow, steep-sided valleys. **Communications** have improved recently with the building of new roads and tunnels through the Alps. Many places however, remain very **isolated**.

The weather here is usually dry and sunny although in winter there is much snow and it can be very cold. Tourism – especially winter sports – has become a major industry in the area. Some people are concerned that the development of winter sports facilities is spoiling the area.

The Industrial Triangle is different altogether. The area is mainly flat and low-lying. It has hot and sunny summers but winters can be wet and foggy.

There are good communications, many towns, and the region is **densely populated**. Although conditions are ideal for farming, it is manufacturing industry that has developed most in recent years. The area has become Italy's industrial heartland and the wealthiest region.

Growth in the Industrial Triangle has been very rapid and there have been difficulties with traffic congestion and pollution in recent years.

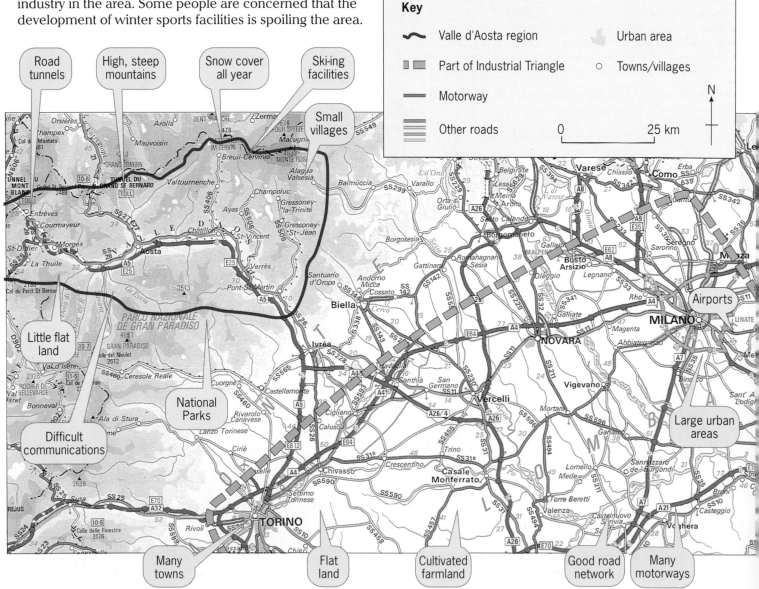

Key

〜 Valle d'Aosta region Urban area

▮ ▭ Part of Industrial Triangle ○ Towns/villages

━ Motorway

▤ Other roads 0 25 km

N

Road tunnels

High, steep mountains

Snow cover all year

Ski-ing facilities

Small villages

Little flat land

Difficult communications

National Parks

Many towns

Flat land

Cultivated farmland

Good road network

Many motorways

Large urban areas

Airports

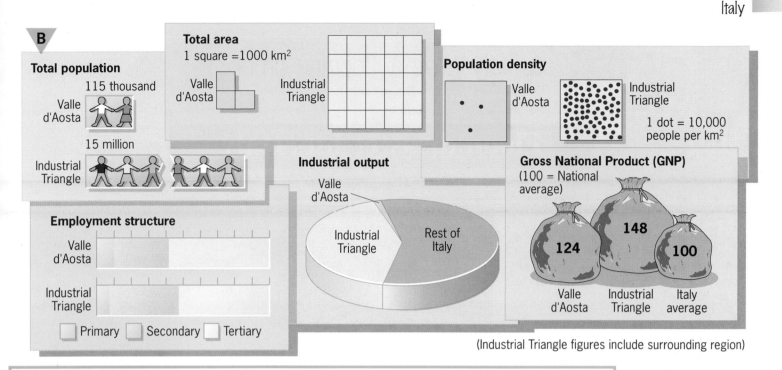

B

Total population

115 thousand
Valle d'Aosta

15 million
Industrial Triangle

Total area
1 square =1000 km²

Valle d'Aosta

Industrial Triangle

Population density

Valle d'Aosta

Industrial Triangle

1 dot = 10,000 people per km²

Industrial output

Valle d'Aosta

Industrial Triangle

Rest of Italy

Gross National Product (GNP)
(100 = National average)

148
124
100

Valle d'Aosta | Industrial Triangle | Italy average

Employment structure

Valle d'Aosta

Industrial Triangle

☐ Primary ☐ Secondary ☐ Tertiary

(Industrial Triangle figures include surrounding region)

Activities

1 Use map **A** to answer these questions.
  a) What features have made industry difficult in the Valle d'Aosta region?
  b) What features help industry in the Industrial Triangle region?
  c) What features in the Valle d'Aosta region do you think would help to attract tourists?

2 a) Make a larger copy of Fact File **C**.
  b) Write the correct word from the bracket in each box.
  c) Give figures for each box.

3 a) Make a copy of table **D**.
  b) Complete the employment structure figures.
  c) Give an example of an industrial activity for each one. Choose from:
  Car making • Ski-ing • Banking and finance • Quarrying • Farming.

4 Explain what the GNP information shows about the wealth of the two regions.

5 Write a decription of each of the two regions.
  • Use the headings given below.
  • Write about 100 words for each one.

• **Physical features** (relief and climate)
• **Human features** (population, towns, communications)
• **Economic features** (industry, jobs)
• **Problems**

C

Fact File

	Valle d'Aosta	Industrial Triangle
Population (most or least)		
Area (largest or smallest)		
How crowded (most or least)		
Industrial output (most or least)		

D

Region	Primary	Secondary	Tertiary
Valle d'Aosta			60%
Example		High-tech	
Industrial Triangle		42%	
Example			

Summary The Valle d'Aosta and Industrial Triangle regions are two important but different regions of Italy.

Does Italy have a flooding problem?

A

Flood disaster hits Italy

Tuesday, November 8 1994

Torrential rain and flooding has once again hit northern Italy. The number of people killed in the area rose to over 100 last night. Thousands more were made homeless as landslides swept houses away.

The weekend storms – the worst in Italy for 80 years – turned rivers into torrents and flattened houses beneath mountains of mud. About 20 000 rescue workers worked through the night near the cities of Turin and Genoa. Road and rail communications were disrupted throughout the north-west.

The town of Asti, 30 miles south-east of Turin and famous for its sparkling wine, suffered some of the worst damage. More than 8000 people were trapped in their homes by flood water and residents spent the night pumping water out of flats, shops and banks. Traffic lights were out of action in the mud-strewn streets and most homes were without electricity, gas or telephones. "We have lost everything," said Maurizio Fiora in his washed out window-frame making business. "It has cost me at least £150 000 and my insurance does not cover this sort of disaster."

Despite the problems in Asti, hundreds more people were brought into the town from the surrounding countryside. They had been plucked from the roofs of their flood-surrounded homes by helicopter. Most slept on the floor of the local school. They were given hot food by the army.

Most problems were caused by the Tanaro and Borbone rivers bursting their banks after some of the worst rainstorms this century. The rain was so heavy that many reservoirs in the region filled to danger levels. In the worst cases, floodgates had to be opened to prevent dams from bursting.

Flood water released from a reservoir at danger-level

Flood damage in Turin

Whilst heavy rain was obviously the main cause of flooding in northern Italy, many people blame developers for making the disaster worse. Adriano Sansa, the Mayor of Genoa, complained that:

66 For decades we have encouraged building programmes and destroyed the forests and countryside without any thought for the environmental effects. In the last 25 years over 8000 square miles of our woodland and countryside has been destroyed. During that time the number of landslides a year have doubled to more than 4000 and 'serious' flooding has become a yearly occurrence. 99

The point that the Mayor is making is very important. The aim of development, whether it be in the form of new ski-ing resorts or new factories, should be to bring about an improvement in conditions for people. In doing that, however, it should not harm or destroy the environment either now or in the future. Development like this is called **sustainable development**.

Sustainable development does not waste resources or damage the environment. It is progress that can go on for year after year. It helps improve our quality of life today but does not spoil our chances in the future.

Since the disaster, there has been much discussion in Italy on how to reduce the damaging effects of development. Most people agree that there is a need for caution and care and much better planning if sustainable development is to be achieved. Whether the steps taken to improve the quality of development are successful or not will be seen in the years ahead.

B How development can increase flooding

More industrial development in the lowlands

Demand for more ski-ing facilities in the mountains

More land needed

Trees chopped down and buildings put up in the countryside

Unprotected soil washed into rivers – rivers fill up with silt

Less rainfall absorbed by vegetation – more water goes into rivers

Rivers overflow and cause flooding

Activities

1 Complete a flood Fact File using the headings shown on the right.

C

Fact File			
Place		Dead	
Date		Homeless	
Main towns		Damage	
Main rivers			

2 Complete these three sentences:
 • Sustainable development is…
 • Sustainable development should…
 • Sustainable development should not…

3 What building developments have there been in northern Italy
 a) in the mountains,
 b) in the lowland areas?

4 Explain the two newspaper headings on the right. Write about 50 words for each one.

Skiers blamed for flood disasters

New industries bring problems not benefits

Summary

Flooding is a serious problem in northern Italy. Rapid and sometimes poorly planned development has been blamed for making the problem worse.

How interdependent is Italy?

When countries work together and rely on each other for help, they are said to be **interdependent**. Being interdependent can help a country progress and improve its standards of living.

One of the main ways that countries become interdependent is by selling goods to each other. They buy things that they need or would like to have. They then sell things to make money to pay for what they have bought. The exchanging of goods and materials like this is called **trade**.

For many years Italy has realised the importance of being a trading nation and has developed links with countries all around the world. Nowadays, as graph **A** shows, her most important trading partners are countries of the **European Union (EU)**.

Italy joined the EU as one of the six founder members in 1957. Since then, membership has grown to fifteen countries and the total market population to over 360 million people. This has provided Italy with a large and accessible market for her imports and exports. It has also enabled the country to benefit from being a member of one of the world's richest and most important groups of nations.

A Italy's trading partners

Italy's trading partners

Imports   Exports

60%			
54%			
	9% 9%	2% 6%	29% 31%
EU countries	USA	Japan	Rest of world

The European Union (EU)   **B**

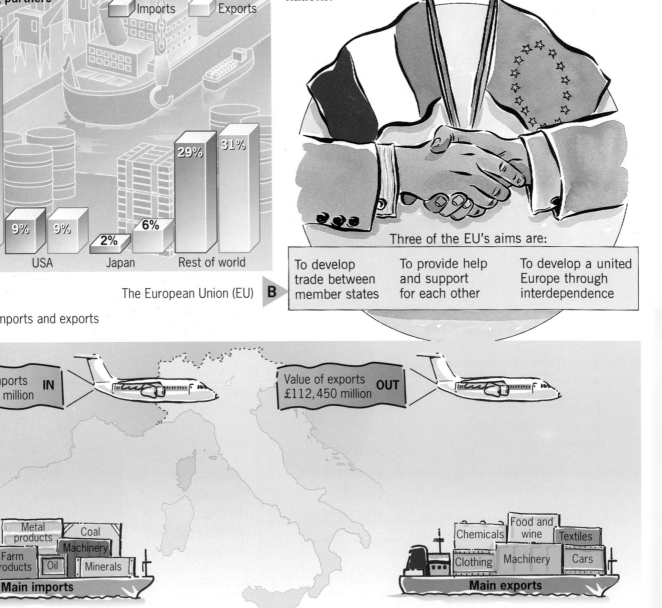

Three of the EU's aims are:

| To develop trade between member states | To provide help and support for each other | To develop a united Europe through interdependence |

C   Italy's imports and exports

Value of imports **IN** £120,070 million

Value of exports **OUT** £112,450 million

Metal products   Coal   Machinery   Farm products   Oil   Minerals

Main imports

Chemicals   Food and wine   Textiles   Clothing   Machinery   Cars

Main exports

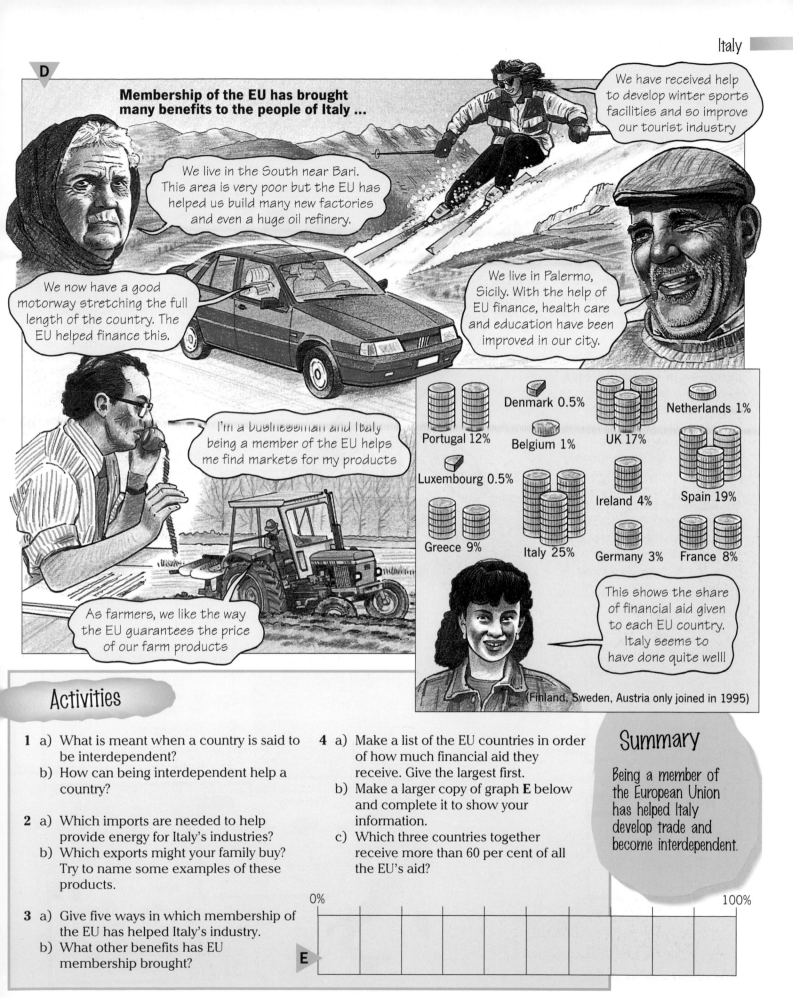

D

Membership of the EU has brought many benefits to the people of Italy ...

We have received help to develop winter sports facilities and so improve our tourist industry

We live in the South near Bari. This area is very poor but the EU has helped us build many new factories and even a huge oil refinery.

We now have a good motorway stretching the full length of the country. The EU helped finance this.

We live in Palermo, Sicily. With the help of EU finance, health care and education have been improved in our city.

I'm a businessman and Italy being a member of the EU helps me find markets for my products

As farmers, we like the way the EU guarantees the price of our farm products

Denmark 0.5%
Portugal 12%
Belgium 1%
UK 17%
Netherlands 1%
Luxembourg 0.5%
Ireland 4%
Spain 19%
Greece 9%
Italy 25%
Germany 3%
France 8%

This shows the share of financial aid given to each EU country. Italy seems to have done quite well!

(Finland, Sweden, Austria only joined in 1995)

Activities

1 a) What is meant when a country is said to be interdependent?
   b) How can being interdependent help a country?

2 a) Which imports are needed to help provide energy for Italy's industries?
   b) Which exports might your family buy? Try to name some examples of these products.

3 a) Give five ways in which membership of the EU has helped Italy's industry.
   b) What other benefits has EU membership brought?

4 a) Make a list of the EU countries in order of how much financial aid they receive. Give the largest first.
   b) Make a larger copy of graph **E** below and complete it to show your information.
   c) Which three countries together receive more than 60 per cent of all the EU's aid?

Summary

Being a member of the European Union has helped Italy develop trade and become interdependent.

0%   100%

E

How developed is Italy?

The study of **development** is important in geography. Development is about progress and improving the quality of life for people. Progress and improvement should not, however, be at the expense of other people and it should not be damaging to the environment. In a word, development should be **sustainable**.

Measuring development is not easy. The most commonly used method is to look at wealth. This is because many people consider that if a country is wealthy, the people living there will be happy, contented and enjoy high **standards of living**. This is not always true. Sometimes a country's wealth does not go towards bettering people's lives, nor is it always spread evenly.

A ▼ Development is about…

More wealth

Longer life expectancy

Better education

Better health

The measurement of wealth by itself, therefore, is not always a good indicator of development. Other factors should also be considered. Some of these are shown in diagram **A**.

Whatever methods are used to measure development, most people agree that Italy is one of the world's most developed countries. Look at diagram **B** below which compares Italy with three other countries in the European Union (EU).

All of these countries are very well-off. As members of the EU they belong to one of the richest and most developed groups of nations in the world. Yet Italy in most cases, is equal to the best of them. Certainly by world standards, Italy is a modern, wealthy and highly developed nation.

B ▼

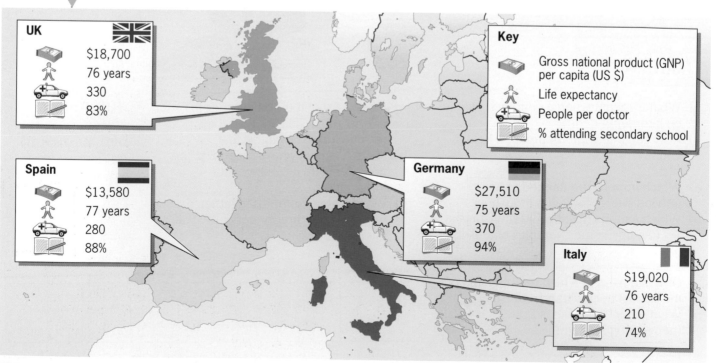

UK

$18,700
76 years
330
83%

Key

Gross national product (GNP) per capita (US $)

Life expectancy

People per doctor

% attending secondary school

Spain

$13,580
77 years
280
88%

Germany

$27,510
75 years
370
94%

Italy

$19,020
76 years
210
74%

The statistics for Italy as a whole, however, hide some problems. Maps **C** and **D** below show that wealth and quality of life are not evenly spread across the whole country. The North is mainly wealthy and has high standards of living. The South is less well developed. Many people there live in difficult conditions and have a much poorer quality of life.

Considerable effort has been put into improving living standards in the South. Over the past 40 years both the EU and the Italian government have invested enormous amounts of money in the region. Some progress has been made but the problem is proving to be a difficult one to solve.

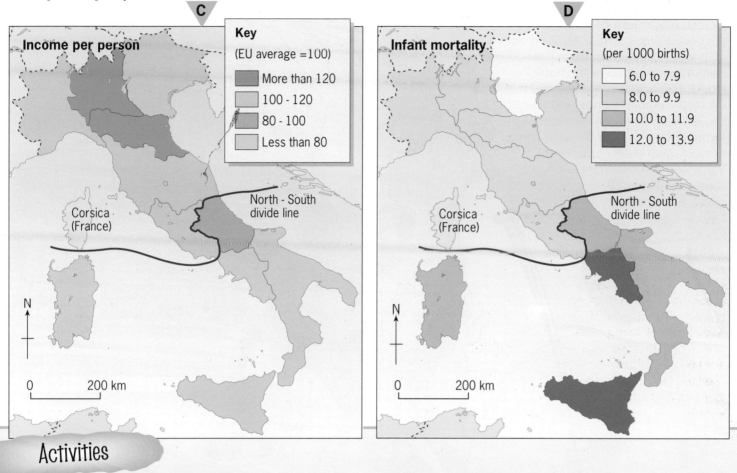

C

Income per person

Key
(EU average =100)

More than 120
100 - 120
80 - 100
Less than 80

North - South divide line

Corsica (France)

N

0   200 km

D

Infant mortality

Key
(per 1000 births)

6.0 to 7.9
8.0 to 9.9
10.0 to 11.9
12.0 to 13.9

North - South divide line

Corsica (France)

N

0   200 km

Activities

1 a) Make a larger copy of table **E** below and complete it using information from diagram **B**.

 b) List the countries from your table in order of
 - GNP/capita (highest first)
 - Life expectancy (highest first)
 - Number of people per doctor (lowest first)
 - % attending secondary schools (highest first)

 c) Look carefully at your completed lists. In your opinion, which two countries are the most developed? Give reasons for your choice.

2 In which part of Italy are
 a) incomes highest,
 b) workers likely to earn most money,
 c) the lowest infant mortality rates,
 d) babies most likely to die early?

3 Write a report expressing your worries about the uneven spread of wealth and living conditions in Italy. Try to include the following.

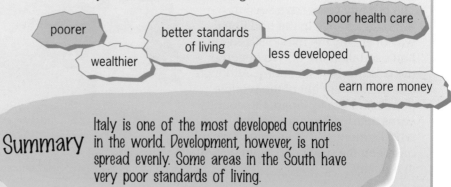

poorer

wealthier

better standards of living

less developed

poor health care

earn more money

E

	UK	Spain	Germany	Italy
GNP/capita				
Life expectancy				
People per doctor				
% attending secondary schools				

Summary Italy is one of the most developed countries in the world. Development, however, is not spread evenly. Some areas in the South have very poor standards of living.

5 Japan

What is Japan like?

Japan consists of four large islands and over 1000 smaller ones. It is located off the east coast of Asia in the Pacific Ocean (map **A**). For much of its history Japan has been isolated from most countries of the outside world. This isolation has allowed it to develop a unique culture and way of life. Recently many characteristics of this traditional way of life have spread to most other parts of the world. Some of these characteristics are shown in diagram **B**.

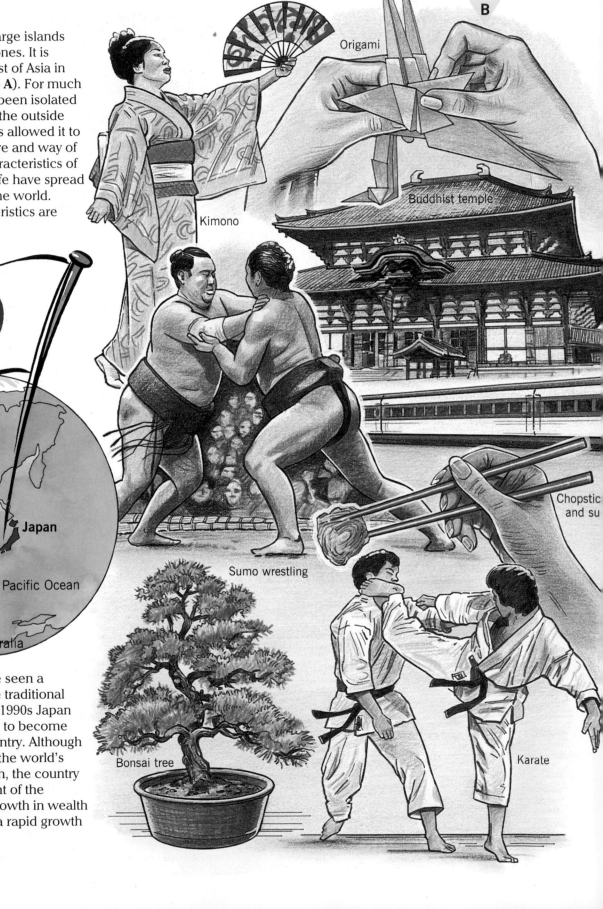

Origami

Kimono

Buddhist temple

A

Asia

Japan

Pacific Ocean

Australia

Sumo wrestling

Chopstic and su

Bonsai tree

Karate

The last fifty years have seen a dramatic change in the traditional way of life. By the mid-1990s Japan had overtaken the USA to become the world's richest country. Although less than 3 per cent of the world's population live in Japan, the country earns nearly 10 per cent of the world's money. This growth in wealth has been the result of a rapid growth in industry.

Most Japanese spend long hours at work and have few holidays. They are very loyal to their place of work and to their family. They are also highly educated and skilled. Leading Japanese companies, like Toyota and Sony, have become household names across the world. It is likely that many things made by these companies can be found in your own home. Some of these goods are shown in diagram **C**.

C

Activities

1 Work with a partner and make a list of things which come to mind when you think about Japan.

2 Diagram **B** shows several features of the traditional Japanese way of life. Following a class discussion, write a short description of each feature.

3 Diagram **C** shows several items which your family might own. These items are made by such Japanese companies as Canon, Casio, Hitachi, Honda, JVC, Mazda, Minolta, Mitsubishi, Nissan, Pentax, Ricoh, Sharp, Toshiba and Toyota. Try to match up the items with the firms which make them (some companies may make more than one item). Put your answers in a table like this one.

D

		Company
Cameras		
Motor vehicles (cars)		
TVs and videos		
Calculators and computers		

Summary

The Japanese live on a group of islands. Their isolation has allowed the development of a unique culture and way of life.

Japan — a land of volcanoes and earthquakes

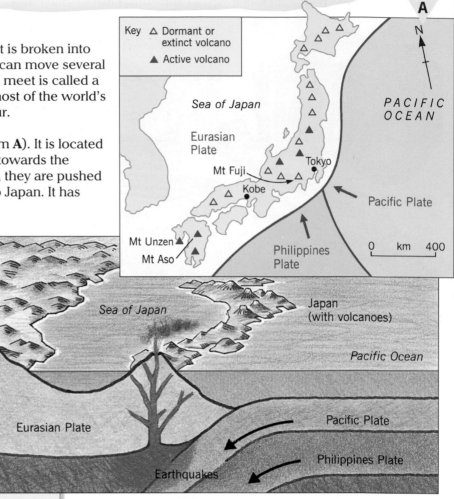

Key
△ Dormant or extinct volcano
▲ Active volcano

Sea of Japan

Eurasian Plate

Mt Fuji

Tokyo

Kobe

Mt Unzen

Mt Aso

PACIFIC OCEAN

Pacific Plate

Philippines Plate

0   km   400

A

The surface of the earth is not all in one piece. It is broken into several large sections called **plates**. Each plate can move several centimetres a year. The place where two plates meet is called a **plate boundary**. It is at plate boundaries that most of the world's major volcanic eruptions and earthquakes occur.

Japan is at the boundary of three plates (diagram **A**). It is located where the Pacific and Philippines Plates move towards the Eurasian Plate. As they meet the Eurasian Plate, they are pushed downwards. This movement has brought **life** to Japan. It has also resulted in **death**.

As part of the crust is pushed downwards, friction causes it to melt. The liquid rock (**magma**) rises to the surface where it forms volcanoes. Without the formation of volcanoes there would be no Japan – the country owes its existence and **life** to volcanoes. Some Japanese volcanoes are still **active**. This means that they still erupt (photo **B**). Occasionally this can cause loss of life (extract **c**). Other volcanoes are either **extinct** and may never erupt again, or they are **dormant** and have not erupted for many years.

Sea of Japan

Japan (with volcanoes)

Pacific Ocean

Eurasian Plate

Pacific Plate

Philippines Plate

Earthquakes

C

B

Mount Unzen, on the southern island of Kyushu, had been dormant for almost 200 years until it erupted violently in June 1991.

Red hot lava rushed down the mountain-side at speeds of up to 120 km/hour. At the same time, clouds of poisonous gas were released into the air.

The eruption destroyed a small village, caused the death of 40 people and left many families homeless.

The last eruption in 1792 killed over 15 000 people.

Plates are not pushed downwards easily or smoothly due to friction. The pushing downwards of a plate often needs considerable force. This force comes from a build up of pressure. If this increase in pressure is released suddenly, then the plate may jerk forwards in a violent movement which is called an earthquake. Of 10 000 earthquakes in the world each year, over 1 000 occur in Japan. The majority are fairly gentle, but occasionally they can mean **death** for thousands of people and can destroy whole cities. An earthquake in 1923 resulted in 140 000 deaths in Tokyo and Yokohama. A more recent one in Kobe in 1995 killed over 5 000 people (photo **D**).

D

E

1.4 million people live in the prosperous port city of Kobe. Although earthquakes are frequently felt in the city, Kobe normally escapes any severe earth movements. That was until 17 January 1995! Early that day, fortunately before the morning rush-hour, an earthquake measuring 7.2 on the **Richter scale** devastated the city. It left over 5 000 dead, 10 000 injured and 250 000 homeless. Although it only lasted 20 seconds, the earthquake caused expressways and many buildings to collapse (photo **D**). Rescue work was hampered because of fears of further damage caused by 'after shocks', the total congestion of the road system, broken water mains, and hundreds of fires (photo **E**). Most fires resulted from broken gas pipes. Many fires lasted for several days, giving the sky an orange glow at night and leaving large areas of burned-out buildings and piles of ash. Many people were forced to camp on pavements in freezing temperatures either because they were homeless or because they were too afraid to go back into their homes.

Activities

1 a) Make a larger copy of diagram **F**.
  b) Add labels to show why volcanic eruptions and earthquakes occur in Japan.

2 Design a front page for a newspaper covering the Kobe earthquake. Try to include the following:
- an eye-catching headline,
- a simple map to show where the earthquake took place,
- a paragraph describing the main effects of the earthquake,
- two short paragraphs giving two eye-witness accounts of the scenes following the earthquake (one could be a night-time description),
- a simple drawing to show one of the effects of the earthquake.

F

J_____

A_____ v_____

M_____ or
e_____ v_____

P_____ O_____

E_____ P____

Rocks melt into m_____

E_____

P_____ P____

Summary

The movement of plates created Japan with its mountains and volcanoes. They also cause volcanic eruptions and earthquakes which can destroy cities and leave many people dead.

Where do people live in Japan?

The population of Japan, as in most other countries, is not evenly spread out. This spread, or **distribution**, of people is shown on map **A**. As 87 per cent of Japan is mountainous, there are large areas where few people live. Most Japanese live in large towns and cities which are crowded onto narrow strips of flat land that lie next to the sea. These coastal areas are among the most **densely** populated regions in the world (photos **B** and **D**). Parts of central Tokyo have the highest population densities in the world (diagram **C**).

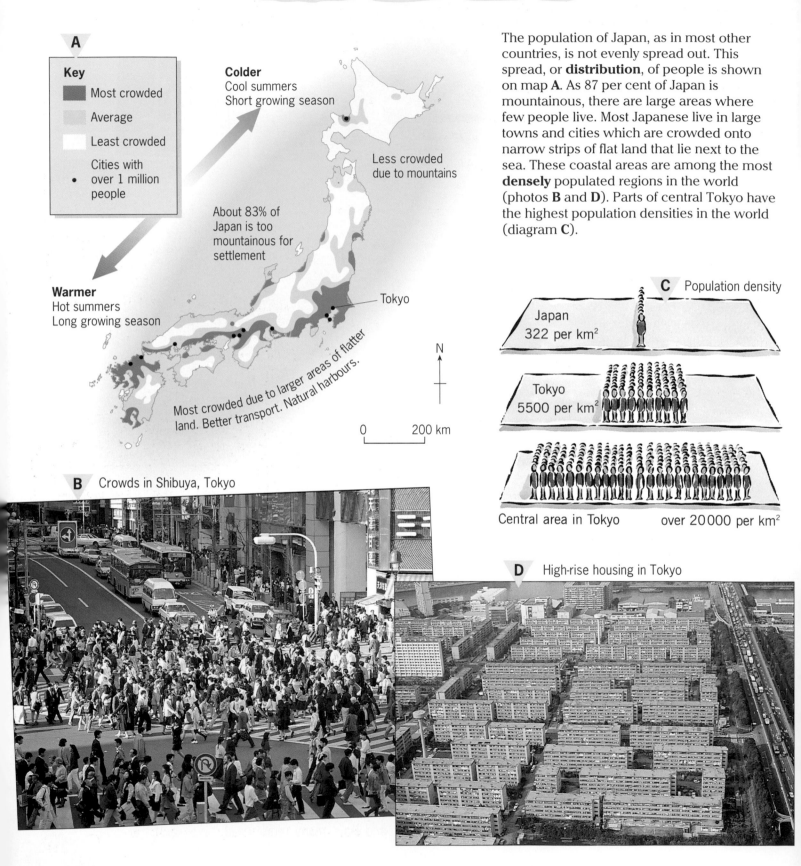

A

Key
- Most crowded
- Average
- Least crowded
- • Cities with over 1 million people

Colder
Cool summers
Short growing season

Less crowded due to mountains

About 83% of Japan is too mountainous for settlement

Warmer
Hot summers
Long growing season

Tokyo

Most crowded due to larger areas of flatter land. Better transport. Natural harbours.

N

0   200 km

C Population density

Japan
322 per km²

Tokyo
5500 per km²

Central area in Tokyo   over 20 000 per km²

B Crowds in Shibuya, Tokyo

D High-rise housing in Tokyo

How is Japan's population changing?

The total population of Japan is increasing very slowly (graph **E**). In fact it is expected to decrease after the year 2025. Graph **F** shows that this slow increase was mainly due to a very low **birth rate** and an almost equally low **death rate**. Once birth rates fall below death rates then there will be a **natural decrease** in the population.

Japan's population problems

Japan's slow and declining increase in population is causing two major problems.

1   The low birth rate has resulted in small families. Small families are favoured because most Japanese houses are small and it is a very expensive country in which to bring up children. However, this means that Japan is becoming increasingly short of people of working age.

2   The falling death rate is producing an **ageing population**. Japanese people have the world's longest **life expectancy** (diagram **G**). This is mainly because they have the money to spend on health care and on healthy diets. Estimates suggest that one-quarter of Japanese people will be over the age of 65 by the year 2020. This is likely to put extra strain on the cost of health care.

E   Japan has a slow rate of population growth

F

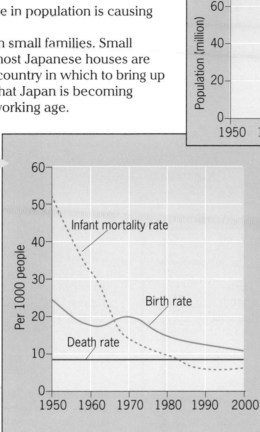

Infant mortality rate

Birth rate

Death rate

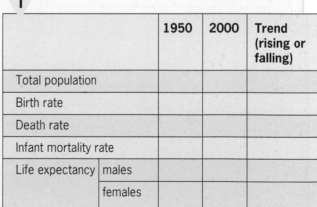

The life expectancy for a Japanese man was 62 in 1950. It is likely to be 78 in 2000

The life expectancy for a Japanese woman in 1950 was 66. It will probably be 84 in 2000

G

Activities

1   a) Which of the islands named on map **H** is the least crowded?
    b) Give three reasons why relatively few people live on this island.
    c) Why do many Japanese crowd into large cities?
    d) On which island are most of the largest cities?
    e) Give three reasons why most large cities are found on this island.

2   a) Make a copy of table **I**. Use the information given in diagrams **E**, **F** and **G** to complete it.
    b) Describe two problems caused by Japan's low birth rate and its increasing life expectancy.

H

Hokkaido

Honshu

Shikoku

Kyushu

I

		1950	2000	Trend (rising or falling)
Total population				
Birth rate				
Death rate				
Infant mortality rate				
Life expectancy	males			
	females			

Summary

Japan's population is uneven in its distribution and is one the slowest growing in the world.

Japan — a land of contrasts

There are many contrasts in Japan's climate, scenery and way of life. The extreme south of Kyushu (map **A**) lies in the same latitude as Cairo in North Africa. This island has active volcanoes, tropical plants and a very warm, wet climate (diagram **B**).

The extreme north of Hokkaido lies in the same latitude as Milan in north Italy. The mountainous island has a cold climate (diagram **C**). Sapporo has an annual snow festival and hosted the 1972 Winter Olympics.

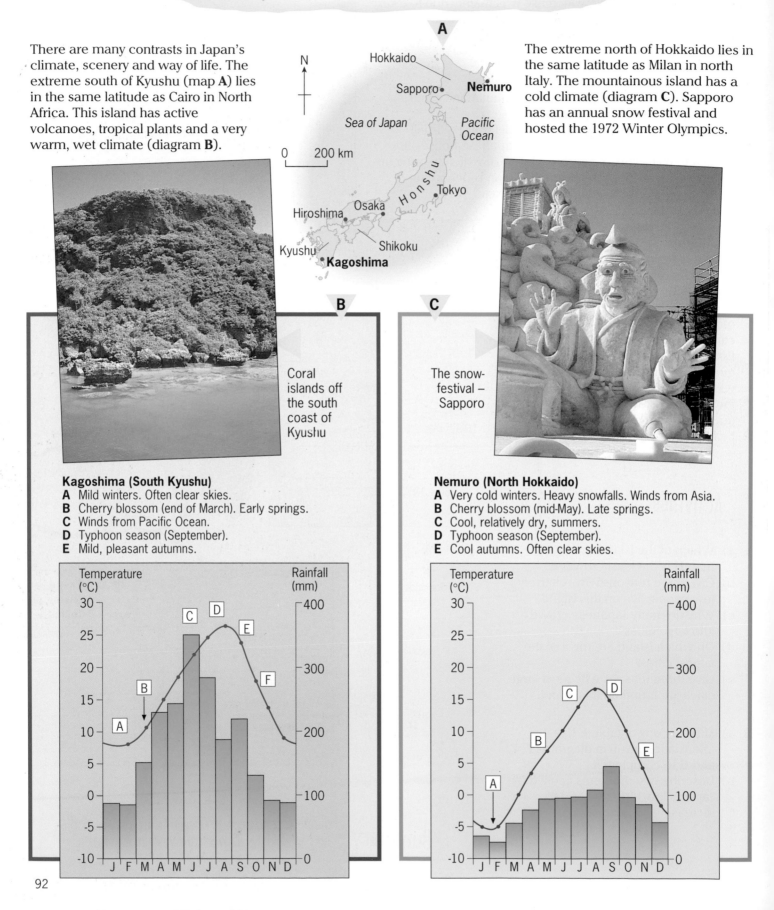

A

N

Hokkaido

Sapporo **Nemuro**

Sea of Japan

Pacific Ocean

0   200 km

Honshu

Tokyo

Osaka

Hiroshima

Kyushu   Shikoku

Kagoshima

B

Coral islands off the south coast of Kyushu

C

The snow-festival – Sapporo

Kagoshima (South Kyushu)
A Mild winters. Often clear skies.
B Cherry blossom (end of March). Early springs.
C Winds from Pacific Ocean.
D Typhoon season (September).
E Mild, pleasant autumns.

Nemuro (North Hokkaido)
A Very cold winters. Heavy snowfalls. Winds from Asia.
B Cherry blossom (mid-May). Late springs.
C Cool, relatively dry, summers.
D Typhoon season (September).
E Cool autumns. Often clear skies.

Honshu, the largest island, lies between Hokkaido and Kyushu. It has high mountains, large forests, many volcanoes and attractive lakes. Its climate is at its best during cherry blossom time in spring and when the leaves of trees change colour in autumn. Honshu is also the most populated island (page 90). It contains many traditional wooden castles, shrines and temples, as well as large modern cities such as Tokyo and Osaka.

D Mount Fuji and Lake Ashi

Itsukushima Shrine on Miyajima Island, near Hiroshima

E

Osaka Castle and high-rise office blocks **F**

G Central Tokyo

Activities

1. a) Copy and complete table **H** to show the differences in climate between Kyushu and Hokkaido.
   b) Why do cherry trees blossom in Kyushu six weeks before they do in Hokkaido?
   c) Give two reasons why Sapporo is a good place for Winter Olympics.

2. Imagine you are spending a few days on the Japanese island of Honshu. Write a postcard home describing
   a) the countryside,
   b) any traditional sites that you might have seen on your visit.

H

		Kyushu	Hokkaido
Temperature	January		
	July		
Rainfall	January		
	July		

Summary Japan is a land of contrasts. There are wide differences in the scenery and climate, and between the traditional and modern way of life.

The growth of Tokyo

Tokyo is one of the world's largest cities. Over 12 million people now live within Tokyo's city boundary, and 32 million within 50 kilometres of the city centre. It has grown so rapidly that is has become joined to several nearby urban areas.

Why was Tokyo a good site for a large settlement?

When we use the word **site** we mean the actual place where a town or city grew up. A site for a settlement was chosen if it had one or more natural advantages. The more natural advantages a place had the more likely it was to grow in size. As diagram **A** shows, Tokyo had many important natural advantages.

What are the functions of present day Tokyo?

For a small settlement to grow into a large city it had to have several specific purposes. Each of these purposes is known as a **function**. As shown in diagram **D**, Tokyo has several functions. It is a major port and industrial centre. It has many high-rise buildings (photo **B**). Most of these belong to big financial firms or are used by the government. Within Tokyo there are several large shopping centres (photo **C**), universities and places of entertainment.

A Growth of Tokyo

Tokyo (pale grey) built on the largest area of flat land in Japan

In early days several rivers (grey green) gave water supply

Much land (pale cream) has been reclaimed from the sea

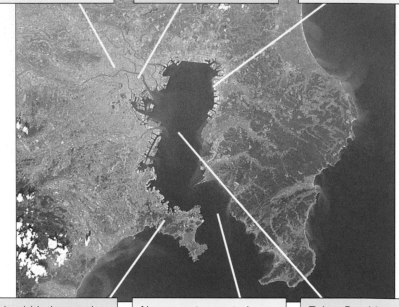

Highland (dark green) – difficult for settlement

Narrow entrance to bay gives protection from severe storms (typhoons)

Tokyo Bay (deep blue) – a deep harbour for large ships

High-rise government buildings **B**

C A busy shopping centre

94

The importance of present day Tokyo **D** ▽

Over 12 million people live in Tokyo itself

Several universities

Tall government office buildings

Centre of business and commerce – e.g. banks, insurance.

Major shopping centres

Major entertainment centre

Many hotels for international conferences

Railways from all over Japan and from surrounding towns meet in Tokyo

The port of Tokyo imports raw materials and sources of energy. It exports manufactured goods.

Industrial centre including hi-tech (videos, cameras) and motor vehicle industries

Reclaimed land

(Note the absence of parks and open space)

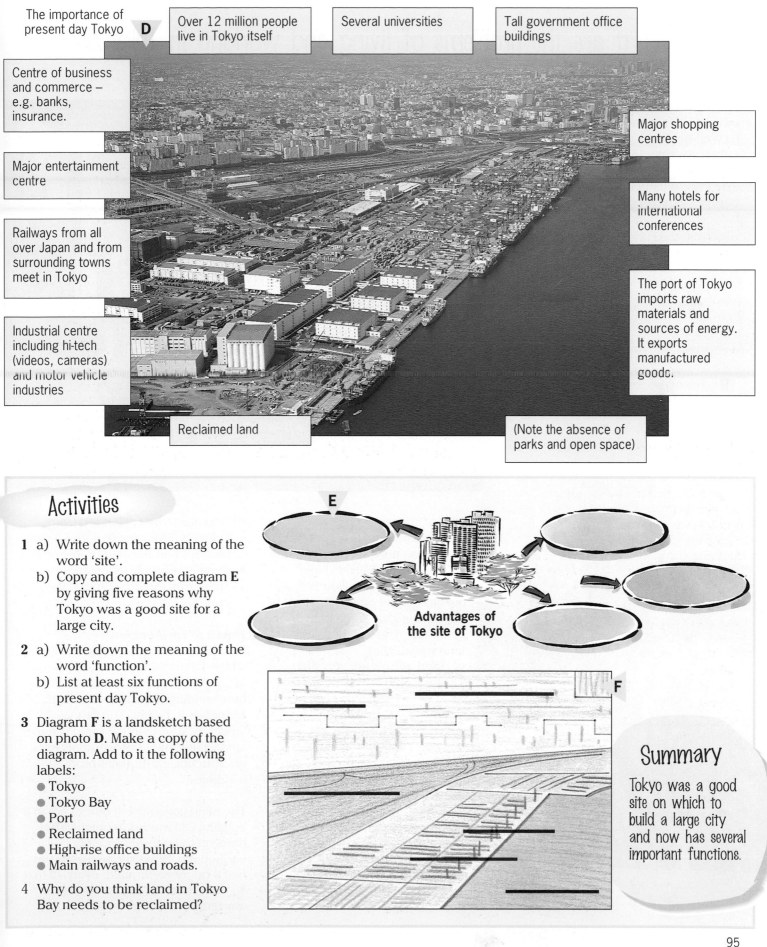

Activities

1 a) Write down the meaning of the word 'site'.

b) Copy and complete diagram **E** by giving five reasons why Tokyo was a good site for a large city.

E

Advantages of the site of Tokyo

2 a) Write down the meaning of the word 'function'.

b) List at least six functions of present day Tokyo.

3 Diagram **F** is a landsketch based on photo **D**. Make a copy of the diagram. Add to it the following labels:
- Tokyo
- Tokyo Bay
- Port
- Reclaimed land
- High-rise office buildings
- Main railways and roads.

F

4 Why do you think land in Tokyo Bay needs to be reclaimed?

Summary

Tokyo was a good site on which to build a large city and now has several important functions.

What are the problems of living and working in Tokyo?

Tokyo is the richest and one of the busiest cities in the world (photo **A**). Its well-paid jobs and exciting life-style have attracted large numbers of people from surrounding areas. However, wealth and the large number of new-comers to Tokyo have created major problems.

'Tokyo is a rapidly growing, rapidly changing city. It is exciting, stimulating and safe. Fast-paced, energetic and busy, the city rarely seems to rest.'
Tourist Guide for Tokyo

A

B The city of Tokyo

Lack of space

Flat land in Japan is very limited. Tokyo was built in one of the few places with a relatively large amount of flat land. However, as the city spread outwards, most of this flat land has been built upon (photo **B**). The lack of usable land means that

- **land values** are very high. High land values mean that it costs a lot of money to buy or rent a piece of land. In parts of Tokyo the price of land can double within a year. Central Tokyo has the highest land values in the world.
- smaller, older buildings have been pulled down and replaced by taller, modern buildings.
- most houses are small in size. Many only have one main room and lack gardens.

- land has to be **reclaimed** from Tokyo Bay for new factories, port facilities and housing (Fact File **D**).
- there is very little **open space**, apart from the area around the Imperial Palace.

Threat of earthquakes

In 1923 an earthquake, and the fires which followed, killed 140 000 people in Tokyo. For many years after this new buildings **had** to be less than 33 metres in height. This law no longer exists due to the rapid rise in land values and improvements in constructing earthquake-proof buildings.

The main government offices are in a building 233 metres in height. They have yet to be tested by a big earthquake. However, as Tokyo has had a major earthquake every 70 to 80 years, another big one is due at any time.

Congestion and pollution

Commuters are people who travel to work. Over 2 million people commute into central Tokyo each day. For many of them the journey can take up to two hours each way. Railway lines have often been built either underground or above existing roads. Railways and expressways become congested at peak times.

Tokyo's road traffic and factories create noise and air pollution. Its inhabitants produce huge amounts of waste. This waste is either burnt or is dumped in Tokyo Bay where it creates more land. Tokyo has several ambitious plans based upon the infilling of parts of Tokyo Bay. One such scheme is described in Fact File **D**.

C Rush hour for Japanese commuters

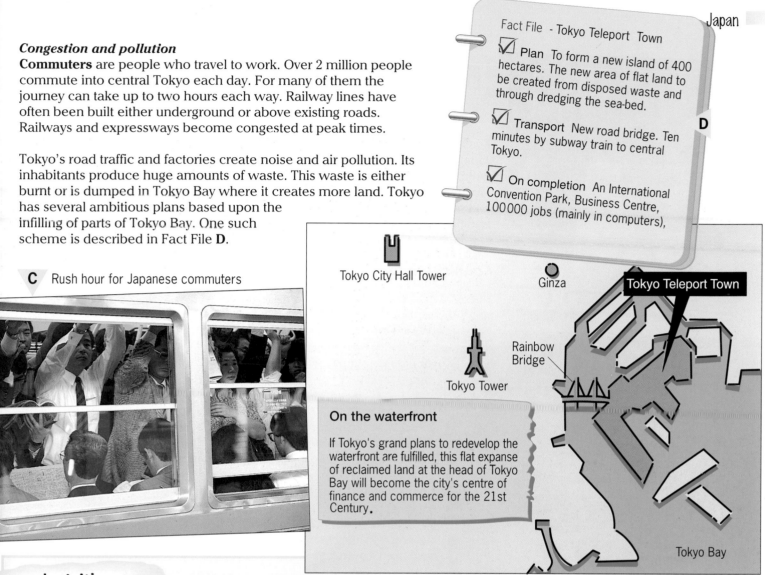

Fact File - Tokyo Teleport Town

☑ **Plan** To form a new island of 400 hectares. The new area of flat land to be created from disposed waste and through dredging the sea-bed.

☑ **Transport** New road bridge. Ten minutes by subway train to central Tokyo.

☑ **On completion** An International Convention Park, Business Centre, 100 000 jobs (mainly in computers),

D

Tokyo City Hall Tower

Ginza

Tokyo Teleport Town

Rainbow Bridge

Tokyo Tower

On the waterfront

If Tokyo's grand plans to redevelop the waterfront are fulfilled, this flat expanse of reclaimed land at the head of Tokyo Bay will become the city's centre of finance and commerce for the 21st Century.

Tokyo Bay

Activities

1 a) What do we mean by the term 'land values'?
   b) Why are land values in Tokyo the highest in the world?

2 Diagram **E** gives the results of a 'Public Opinion Survey on Urban Improvement in Tokyo'. Describe the **causes** of each of the five types of problem referred to in the survey.

3 Describe how the Teleport Town scheme aims to reduce Tokyo's problems of
   a) shortage of land,
   b) travel (commuting) to work,
   c) lack of housing.
   What type of jobs does the scheme hope to create?

E Public opinion survey on urban improvement in Tokyo

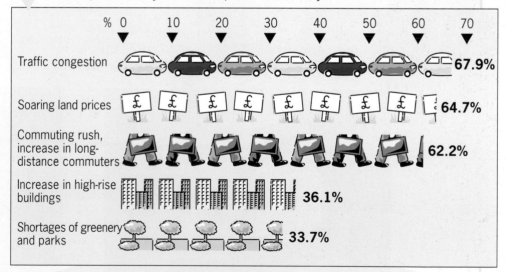

	% 0	10	20	30	40	50	60	70
Traffic congestion								67.9%
Soaring land prices								64.7%
Commuting rush, increase in long-distance commuters								62.2%
Increase in high-rise buildings					36.1%			
Shortages of greenery and parks				33.7%				

Summary

Tokyo's rapid growth has created many problems. These include high land values, congestion, pollution and a lack of open space.

Life in rural Kyushu

Kyushu is the warmest of the main Japanese islands. It is also the one which is furthest away from Tokyo. Much of Kyushu is still mainly rural and many people still live in traditional houses (diagram **A**).

Farming

Akira and Chika, like many other people in rural Kyushu, live on a farm.

A

Konnichiwa, ogenki desula!

That is Japanese for 'Hello. How are you'. Our names are Akira and Chika. We live on a farm in Kyushu. Our house there is typical of many still found in rural areas of Japan.

The outside walls are made of wood while the roof is usually made either from tiles or thatch. The entrance hall has a wooden floor. We have to remove our shoes in the hallway and put on slippers before we enter the rest of the house. Apart from the kitchen, the floors of all our rooms are covered in **tatami**. Tatami are mats made of woven rushes. The inside walls of most rooms are usually sliding doors which we call **shoji**. They can be opened to make a larger room or to let in fresh air during the hot summers.

At meal times we sit cross-legged on floor cushions at a low table. Our main meal is in the evening and it often consists of several courses. Before we go to bed the room has to be cleared. We then lay large mattresses, or **futons**, on the floor ready for us to sleep on. In the morning these are rolled up and stored in large cupboards. Because many houses are small, we often have to use the same room for living, eating and sleeping. Fortunately, few Japanese families have more than two children and so our small houses do not seem to be too overcrowded.

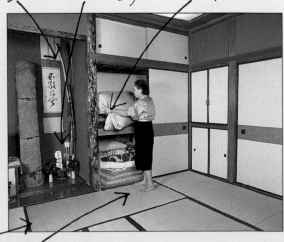

kakejiku (hanging scrolls)

butsudan (family shrine)

Futons stored away during daytime

shoji (sliding doors)

Tatami (matting)

No shoes. Removed on entry to the house.

They are expected to help their parents at busy times of the year. However, like many others of their age, Akira and Chika want to find work in a city and do not want to have their own farm.

We have already seen that the amount of flat land in Japan is limited. Most farms are therefore small in size and have to be farmed **intensively**. Intensive farming is when the maximum use is made of the land and no space is wasted. By using fertiliser and modern machinery

Japanese farmers are able to grow as much as possible on each hectare of land. The warm, wet Kyushu climate also allows two crops to be grown in the same field every year. Rice has always been the most important crop in Japan (diagram **B**). However, the Japanese are now eating a wider diet, especially Western food. As a result rice production has fallen considerably. As most farmers grow less rice they now grow more fruit and vegetables.

B | Rice Growing

1 Cultivators are used to plough the land

2 Tractors prepare the paddies (rice fields)

3 Seedlings are grown in greenhouses

4 Seedlings are planted in the paddies

5 Herbicides and fertilisers are spread

6 Binders harvest the rice

Rice fields

Harvesting rice

Activities

1 Imagine you are either Akira or Chika. Write a letter to a friend who lives in Tokyo. In your letter describe your house and explain how you try to save space within it.

2 a) What is meant by the term 'intensive farming'?
   b) Why is farming in Kyushu intensive?

3 a) Make a larger copy of diagram **C**.
   b) Complete it using information given in diagram **B**.
   c) Add the title 'Rice growing in Kyushu'.

C

Intensive farming in Kyushu

Summary

Most houses and farms in Kyushu, as elsewhere in rural Japan, are small in size. This means maximum use has to be made of space in the house and land on the farm.

Recent changes in Kyushu

Industry began to develop rapidly in Japan after 1950. The main industrial areas were around Tokyo and other large cities on the main island of Honshu. Places furthest away from Tokyo, like Kyushu, tended not to become industrialised. Even as late as the 1980s many people living on Kyushu were still farmers.

Since 1990 there has been a decline, or a **recession**, in industry across the world. Even Japan has been affected by this recession. Yet at this same time industry has developed rapidly in Kyushu causing a decline in the importance of farming. Diagram **A** gives some of the many reasons why industry has been attracted to Kyushu. For example,

- during the 1980s many of Japan's largest electronic firms built factories there;
- in the early 1990s car firms (Toyota and Nissan) opened assembly plants on the island.

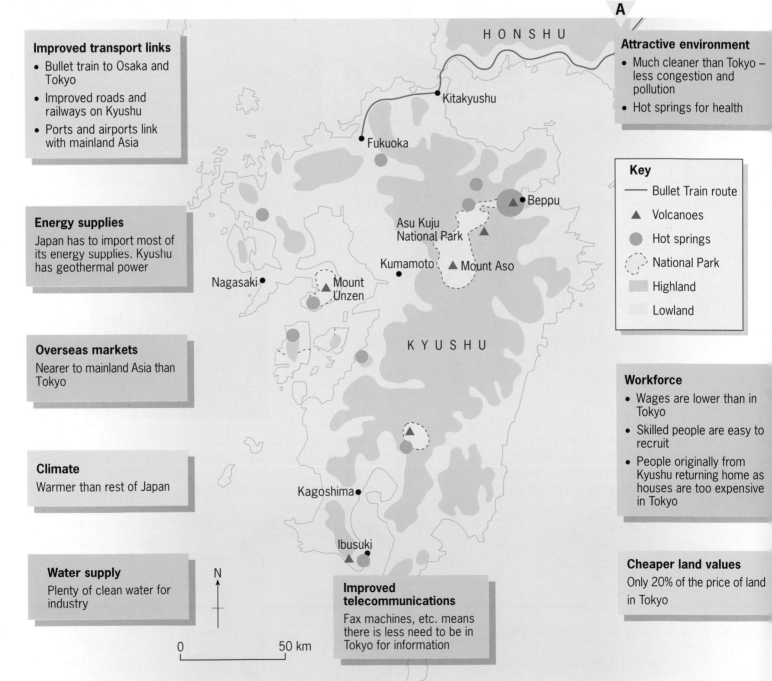

A

Improved transport links
- Bullet train to Osaka and Tokyo
- Improved roads and railways on Kyushu
- Ports and airports link with mainland Asia

Energy supplies
Japan has to import most of its energy supplies. Kyushu has geothermal power

Overseas markets
Nearer to mainland Asia than Tokyo

Climate
Warmer than rest of Japan

Water supply
Plenty of clean water for industry

Attractive environment
- Much cleaner than Tokyo – less congestion and pollution
- Hot springs for health

Workforce
- Wages are lower than in Tokyo
- Skilled people are easy to recruit
- People originally from Kyushu returning home as houses are too expensive in Tokyo

Cheaper land values
Only 20% of the price of land in Tokyo

Improved telecommunications
Fax machines, etc. means there is less need to be in Tokyo for information

Key
— Bullet Train route
▲ Volcanoes
● Hot springs
National Park
Highland
Lowland

HONSHU
Kitakyushu
Fukuoka
Beppu
Asu Kuju National Park
Kumamoto
Mount Aso
Nagasaki
Mount Unzen
KYUSHU
Kagoshima
Ibusuki

N

0   50 km

B Geothermal steam at Beppu

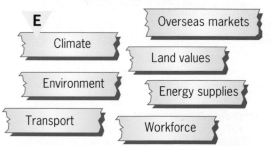

C A hot sand bath at Ibusuki

A second change has been the growth of tourism. Diagram **D** suggests why people living and working in Tokyo might want to visit Kyushu.

D

I want to see the active volcanoes in the National Parks

The air is cleaner and the sky clearer than in Tokyo

I want to visit the health spas and bathe in the hot sands

I want to see the cherry blossom in early Spring and the changing leaves in Autumn

The climate is warmer in winter and more healthy all year than in Tokyo

Activities

1 Look at the captions in diagram **E**. Explain how each caption has helped to attract industry to Kyushu.

E

Climate	Overseas markets
Environment	Land values
Transport	Energy supplies
	Workforce

2 Imagine you are having a short holiday in Kyushu. Write a postcard to a friend in Tokyo. Describe the weather, things you have seen and things you have done during your visit.

3 List some of the ways in which you think that living in Kyushu today is different to living there twenty years ago. You may need to refer to pages 98 to 101 to answer this question.

Summary

Kyushu has recently attracted many new industries and an increasing number of tourists.

Two regions compared

As we have seen, there are considerable differences between Tokyo and Kyushu. The Tokyo region is almost totally built up. It is an urban area with very little open space. Kyushu, in contrast, is mainly rural. There are, however, several cities in the north of the island which are now growing rapidly. As diagram **A** shows, for every person living in one square kilometre in Kyushu, there are 17.5 in Tokyo.

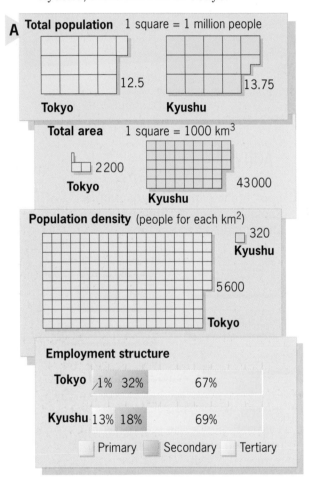

A

Total population　1 square = 1 million people

12.5

Tokyo

13.75

Kyushu

Total area　1 square = 1000 km³

2200

Tokyo

43000

Kyushu

Population density (people for each km²)

320

Kyushu

5600

Tokyo

Employment structure

Tokyo　1% 32% 67%

Kyushu　13% 18% 69%

☐ Primary　☐ Secondary　☐ Tertiary

B

Childhood years	Marriage

1 In Tokyo

With their father at work and rarely home, city children tend to spend most of their time with their mother. In homes in which both parents work, some children must look after the house until their parents come home.

Big cities are full of paved roads and have few parks. Few children have anywhere outside to play. As a result, video games and other indoor amusements are popular.

It is expensive to support a family in the big cities, where the prices of houses, food and clothes are high. The Japanese are having to marry later. In Tokyo it is usual for both husband and wife to have jobs.

2 In Kyushu

Because the parents in a farming household are extremely busy, the children are normally raised by the grandparents.

In farming communities, located in a rich natural environment, children swim in rivers, collect insects and amuse themselves. Few will stay closed up indoors.

The problem in Kyushu is a shortage of young women. This is because many move to big cities to study or to find work.

Activities

1 a) Make a larger copy of Fact File **C**.
　b) Write the correct word or number from the bracket in each box.
　c) Using your completed Fact File, write five sentences to show differences between the two regions.

C

Fact File

		Tokyo	Kyushu
Population	(most or least) (12.5 or 13.75)	.... million	.... million
Area	(larger or smaller) (2 200 or 43 000)	.... km²	.... km²
Population density	(higher or lower) (5 600 or 320)	.... per km²	.... per km²
Primary jobs	(more or fewer) (1 or 13)	.... %	.... %
Secondary jobs	(more or fewer) (32 or 18)	.... %	.... %

The way of life in each region is very different (diagram **B**). Tokyo has many high-rise buildings, office blocks, factories, shops, places of entertainment, roads and railways. The people who live there have one of the highest standards of living in the world. Despite recent industrial growth, Kyushu has relatively few large urban areas. It has mountains, active volcanoes, hot springs, an attractive coastline and several National Parks. Although its standard of living is lower than Tokyo, it is still high compared with most other places in the world.

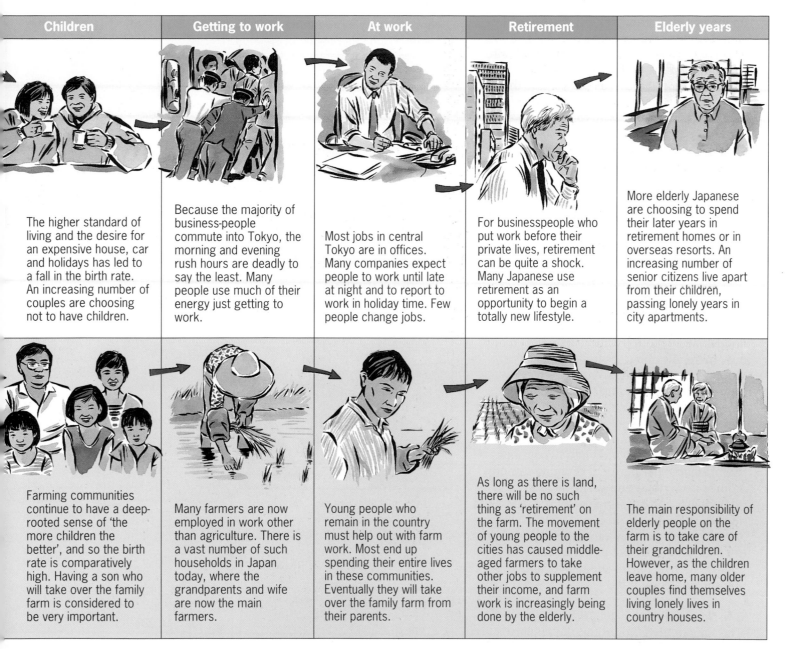

Children	Getting to work	At work	Retirement	Elderly years
The higher standard of living and the desire for an expensive house, car and holidays has led to a fall in the birth rate. An increasing number of couples are choosing not to have children.	Because the majority of business-people commute into Tokyo, the morning and evening rush hours are deadly to say the least. Many people use much of their energy just getting to work.	Most jobs in central Tokyo are in offices. Many companies expect people to work until late at night and to report to work in holiday time. Few people change jobs.	For businesspeople who put work before their private lives, retirement can be quite a shock. Many Japanese use retirement as an opportunity to begin a totally new lifestyle.	More elderly Japanese are choosing to spend their later years in retirement homes or in overseas resorts. An increasing number of senior citizens live apart from their children, passing lonely years in city apartments.
Farming communities continue to have a deep-rooted sense of 'the more children the better', and so the birth rate is comparatively high. Having a son who will take over the family farm is considered to be very important.	Many farmers are now employed in work other than agriculture. There is a vast number of such households in Japan today, where the grandparents and wife are now the main farmers.	Young people who remain in the country must help out with farm work. Most end up spending their entire lives in these communities. Eventually they will take over the family farm from their parents.	As long as there is land, there will be no such thing as 'retirement' on the farm. The movement of young people to the cities has caused middle-aged farmers to take other jobs to supplement their income, and farm work is increasingly being done by the elderly.	The main responsibility of elderly people on the farm is to take care of their grandchildren. However, as the children leave home, many older couples find themselves living lonely lives in country houses.

2 Describe the differences in the way of life between the following groups of people living in Tokyo and Kyushu.
   a) Children (under 15 years)
   b) Young couples (20–30 years)
   c) Middle aged people (30–60 years)
   d) Elderly people (over 60 years)

3 Imagine that you are a writer for a geographical magazine. Write two short articles of about 100 words each to describe the main features of
   a) Tokyo
   b) Kyushu.
   You should refer to any information given in this unit.

Summary

Living and working in Tokyo is very different to living and working in Kyushu.

Sustainable development in Japan

In the last few years the term **sustainable development** has become popular in geography. It is not an easy term to explain. If you refer back to diagram **A** on page 56 you will see that different people use the term in different ways. Ideally, sustainable development should lead to an improvement in people's

- **quality of life** – how content they are with their way of life;
- **standard of living** – how well-off they are economically.

These improvements should be achieved without wasting the earth's resources or destroying the environment. They should not, therefore, just be of benefit to people living today, but also for future generations.

Energy supplies in Japan

Japan has become the world's richest industrialised country. Industry needs many raw materials and uses large supplies of energy. Strangely, Japan has very few raw materials of its own and has very limited energy supplies. This means that it has to rely upon other countries to provide **natural resources**. As diagram **A** shows, natural resources can be divided into two groups. Unfortunately, 87 per cent of Japan's energy comes from **non-renewable** sources, and that means non-sustainable sources. It would appear desirable, from a sustainable development point of view, that Japan should try to develop more of its **renewable** forms of energy such as **hydro-electricity** and **geothermal** (graph **B**).

The Nagara River

The Nagara River flows into the Pacific Ocean to the west of Nagoya (map **C**). In 1990, it was one of the last free-flowing rivers in Japan. Since then a construction company has been building a huge dam across the river. The dam is over 600 metres wide. It will create a huge reservoir of fresh water for nearby cities and industries. Although in many parts of the world the storage of water is seen as an example of sustainable development, the Nagara River scheme has become a major controversy within Japan.

Non-renewable resources are those which can only be used once. Non-renewable resources include coal, oil and natural gas. Many pollute the environment. In time these resources will run out. They are, therefore, non-sustainable.

Renewable resources can be used over and over again. Renewable resources include hydro-electricity and geothermal power. As they will not run out or pollute the environment, they are sustainable.

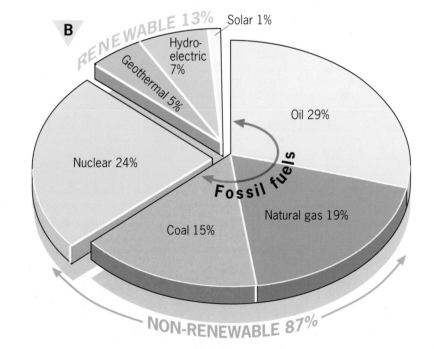

B

RENEWABLE 13%

Solar 1%

Hydro-electric 7%

Geothermal 5%

Oil 29%

Nuclear 24%

Fossil fuels

Natural gas 19%

Coal 15%

NON-RENEWABLE 87%

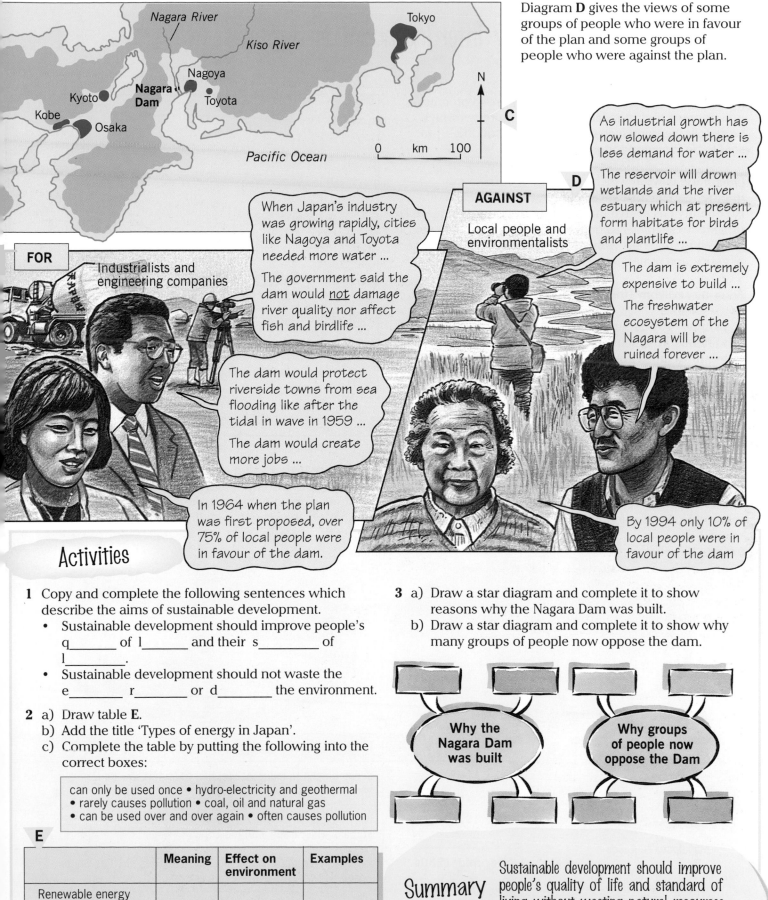

Diagram **D** gives the views of some groups of people who were in favour of the plan and some groups of people who were against the plan.

Activities

1 Copy and complete the following sentences which describe the aims of sustainable development.
   • Sustainable development should improve people's q_____ of l_____ and their s_____ of l_____.
   • Sustainable development should not waste the e_____ r_____ or d_____ the environment.

2 a) Draw table **E**.
   b) Add the title 'Types of energy in Japan'.
   c) Complete the table by putting the following into the correct boxes:

   can only be used once • hydro-electricity and geothermal • rarely causes pollution • coal, oil and natural gas • can be used over and over again • often causes pollution

E

	Meaning	Effect on environment	Examples
Renewable energy			
Non-renewable energy			

3 a) Draw a star diagram and complete it to show reasons why the Nagara Dam was built.
   b) Draw a star diagram and complete it to show why many groups of people now oppose the dam.

Why the Nagara Dam was built

Why groups of people now oppose the Dam

Summary

Sustainable development should improve people's quality of life and standard of living without wasting natural resources or destroying the environment.

105

How interdependent is Japan?

Countries need to work together if they are to progress and improve their standards of living. Countries which work together, or which rely upon other countries, are said to be **interdependent**.

One of the ways in which countries become interdependent is by selling and buying from each other. No country has everything that its people want or need. To provide these things a country will have to exchange goods and materials with other countries. This exchange of goods and materials is called **trade**.

A country will buy (**import**) things which it lacks. These may either be
- **primary goods**, such as foodstuffs and minerals, which are often low in value, or
- **manufactured goods**, such as machinery, which are high in value.

In order to pay for these a country has to sell (**export**) goods of which it has a surplus. Ideally a country aims to have a **trade surplus**. This means that it earns more money from its exports than it spends on its imports.

Japan's trade surplus is one of the biggest in the world (graph **A**). Its trade is typical of many economically more developed countries (graph **B**). Most of its imports are relatively low value foodstuffs and minerals. Most of its

exports are higher value manufactured goods. Japan is therefore able to loan money to, and invest in, other countries. This all helps to make Japan richer. Table **C** summarises the growth and direction of Japan's trade.

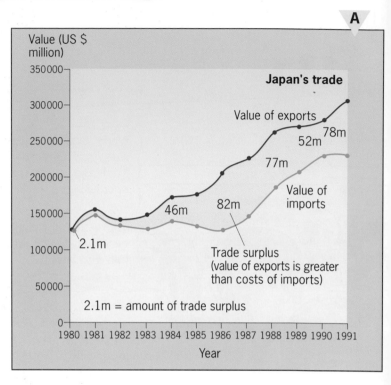

A

Japan's trade

Value of exports

Value of imports

Trade surplus (value of exports is greater than costs of imports)

2.1m = amount of trade surplus

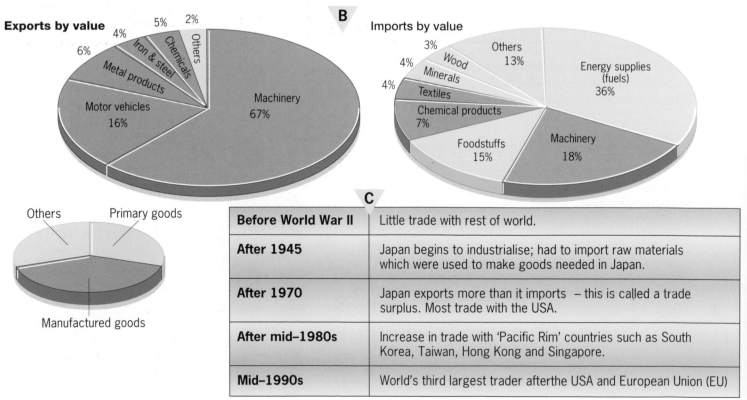

B

Exports by value

Machinery 67%
Motor vehicles 16%
Metal products 6%
Iron & steel 4%
Chemicals 5%
Others 2%

Imports by value

Energy supplies (fuels) 36%
Machinery 18%
Foodstuffs 15%
Chemical products 7%
Textiles 4%
Minerals 4%
Wood 3%
Others 13%

Others
Primary goods
Manufactured goods

C

Before World War II	Little trade with rest of world.
After 1945	Japan begins to industrialise; had to import raw materials which were used to make goods needed in Japan.
After 1970	Japan exports more than it imports – this is called a trade surplus. Most trade with the USA.
After mid–1980s	Increase in trade with 'Pacific Rim' countries such as South Korea, Taiwan, Hong Kong and Singapore.
Mid–1990s	World's third largest trader after the USA and European Union (EU).

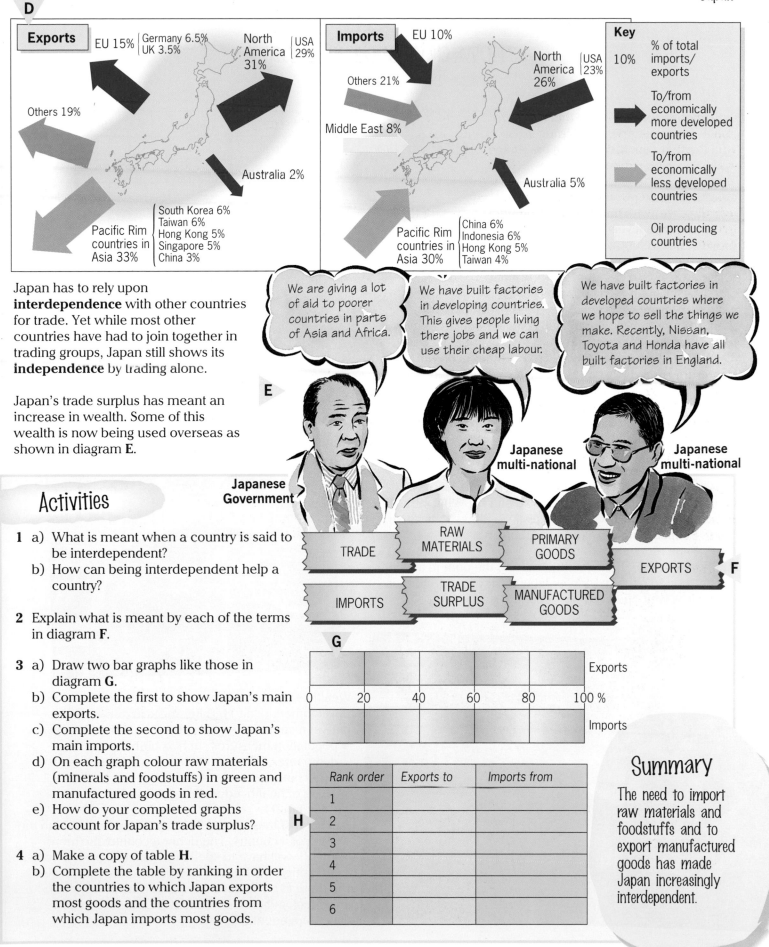

D **Exports**

EU 15% | Germany 6.5% / UK 3.5%

North America 31% | USA 29%

Others 19%

Australia 2%

Pacific Rim countries in Asia 33% | South Korea 6% / Taiwan 6% / Hong Kong 5% / Singapore 5% / China 3%

Imports

EU 10%

North America 26% | USA 23%

Others 21%

Middle East 8%

Australia 5%

Pacific Rim countries in Asia 30% | China 6% / Indonesia 6% / Hong Kong 5% / Taiwan 4%

Key

10% % of total imports/exports

To/from economically more developed countries

To/from economically less developed countries

Oil producing countries

Japan has to rely upon **interdependence** with other countries for trade. Yet while most other countries have had to join together in trading groups, Japan still shows its **independence** by trading alone.

Japan's trade surplus has meant an increase in wealth. Some of this wealth is now being used overseas as shown in diagram **E**.

E

We are giving a lot of aid to poorer countries in parts of Asia and Africa.

We have built factories in developing countries. This gives people living there jobs and we can use their cheap labour.

We have built factories in developed countries where we hope to sell the things we make. Recently, Nissan, Toyota and Honda have all built factories in England.

Japanese Government

Japanese multi-national

Japanese multi-national

F

TRADE — RAW MATERIALS — PRIMARY GOODS — EXPORTS

IMPORTS — TRADE SURPLUS — MANUFACTURED GOODS

Activities

1 a) What is meant when a country is said to be interdependent?
b) How can being interdependent help a country?

2 Explain what is meant by each of the terms in diagram **F**.

3 a) Draw two bar graphs like those in diagram **G**.
b) Complete the first to show Japan's main exports.
c) Complete the second to show Japan's main imports.
d) On each graph colour raw materials (minerals and foodstuffs) in green and manufactured goods in red.
e) How do your completed graphs account for Japan's trade surplus?

4 a) Make a copy of table **H**.
b) Complete the table by ranking in order the countries to which Japan exports most goods and the countries from which Japan imports most goods.

G

Exports

0   20   40   60   80   100 %

Imports

H

Rank order	Exports to	Imports from
1		
2		
3		
4		
5		
6		

Summary

The need to import raw materials and foodstuffs and to export manufactured goods has made Japan increasingly interdependent.

How developed is Japan?

You should now be aware that there are many differences between living in Japan and in other countries such as the United Kingdom. These differences may include ethnic backgrounds, dress, housing, religion, language, jobs and wealth. You will also be aware that different countries are at different levels of **development**. However, development is not an easy term to define. To many people development means how rich a country is and how high a standard of living it has. To make comparisons between countries easier, the wealth of a country is given by its **gross national product (GNP)**. Remember that GNP per person is the amount of money earned by a country divided by its total population. Based upon wealth, countries are said to be either **economically more developed** or **economically less developed**.

	A		
Country	**GNP (US$)**	**Country**	**GNP (US$)**
Japan (Richest)	39 640	Brazil	3 640
USA	26 980	Kenya	280
Italy	19 020	India	340
UK	18 700	Mozambique	80 (Poorest)

Table **A** gives the GNP for several countries including those referred to in this book. Because Japan has the highest GNP it could, therefore, be described as the richest country and the one which is the most economically developed. Certainly, extremely high levels of wealth and technology can be seen in both Japanese industry and every day life (photos **B** and **C**). What table **A** does not show, are the differences in wealth within Japan itself. People living in Tokyo and other large cities are likely to be much richer than those living in rural areas. However, the gap between the rich and the poor in a well-off developed country such as Japan is never as great as it is in a less well-off developing country like Kenya.

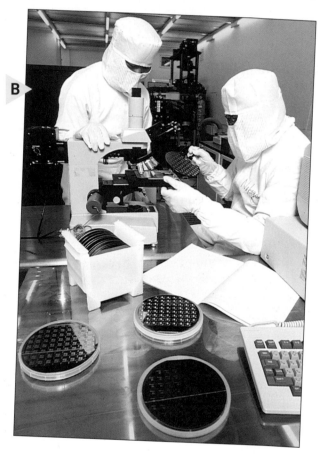
B

C

Wealth is not the only way to measure (or describe) development (Unit 1). Differences in development between countries can also be seen by looking at
● population measures such as birth rates, death rates, infant mortality rates and life expectancy;
● levels of health, education, jobs and trade (diagram **D**).
Yet each of these measures can in turn be linked to the wealth of a country. The richer a country the more money it will have to spend on such things as schools, universities, health care and birth control. It will also be more likely to have a modern transport system, high-tech industries, and to have become more interdependent.

D

Jobs	Trade	Population	Health	Education
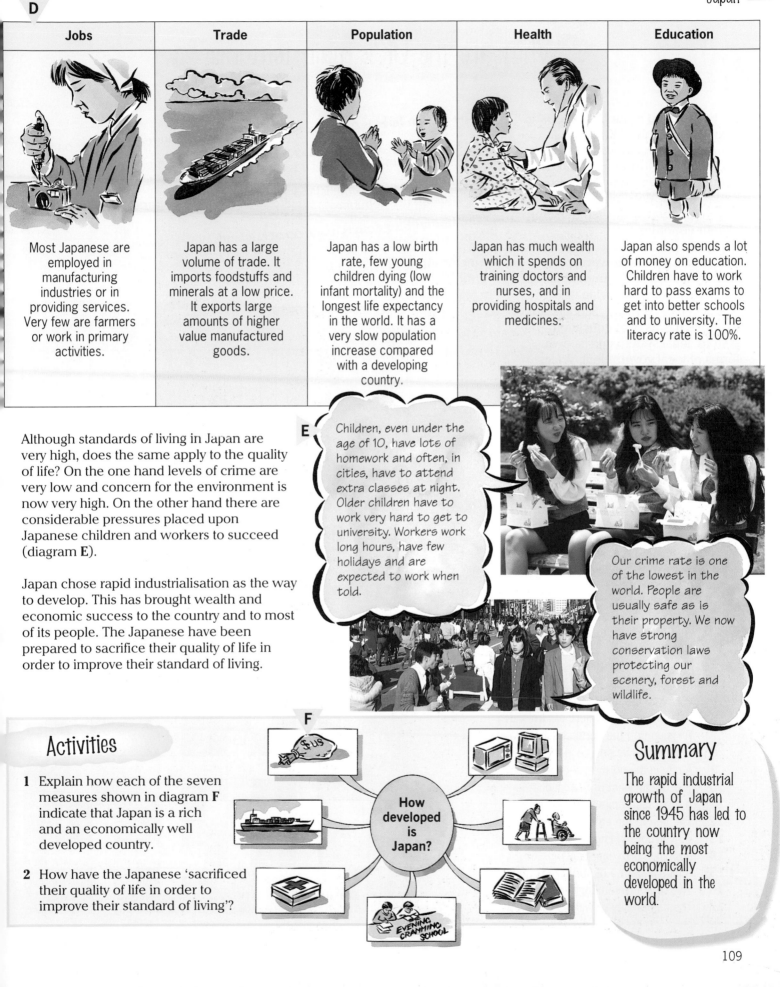				
Most Japanese are employed in manufacturing industries or in providing services. Very few are farmers or work in primary activities.	Japan has a large volume of trade. It imports foodstuffs and minerals at a low price. It exports large amounts of higher value manufactured goods.	Japan has a low birth rate, few young children dying (low infant mortality) and the longest life expectancy in the world. It has a very slow population increase compared with a developing country.	Japan has much wealth which it spends on training doctors and nurses, and in providing hospitals and medicines.	Japan also spends a lot of money on education. Children have to work hard to pass exams to get into better schools and to university. The literacy rate is 100%.

Although standards of living in Japan are very high, does the same apply to the quality of life? On the one hand levels of crime are very low and concern for the environment is now very high. On the other hand there are considerable pressures placed upon Japanese children and workers to succeed (diagram **E**).

Japan chose rapid industrialisation as the way to develop. This has brought wealth and economic success to the country and to most of its people. The Japanese have been prepared to sacrifice their quality of life in order to improve their standard of living.

E

Children, even under the age of 10, have lots of homework and often, in cities, have to attend extra classes at night. Older children have to work very hard to get to university. Workers work long hours, have few holidays and are expected to work when told.

Our crime rate is one of the lowest in the world. People are usually safe as is their property. We now have strong conservation laws protecting our scenery, forest and wildlife.

F

Activities

1 Explain how each of the seven measures shown in diagram **F** indicate that Japan is a rich and an economically well developed country.

2 How have the Japanese 'sacrificed their quality of life in order to improve their standard of living'?

How developed is Japan?

Summary

The rapid industrial growth of Japan since 1945 has led to the country now being the most economically developed in the world.

EVENING CRAMMING SCHOOL

What are the UK's main features?

First of all what is the **UK**? The UK is short for the **United Kingdom** a country made up of England, Wales, Scotland and Northern Ireland. This country is governed by the Parliament at Westminster in London which is attended by members from all four countries.

Within this system however, each country has a certain amount of self rule and is able to develop its own distinct characteristics. Scotland for example has its own education system, legal system and its own forms of local government. Wales has its own language and Northern Ireland separate laws and a distinct system of education.

A

The British Isles is made up of two large islands and many smaller ones.

Great Britain is the largest of the islands. It is sometimes just called Britain. It is divided into three parts – England, Wales and Scotland.

Ireland is the other large island. It is divided into two parts – Northern Ireland and the Republic of Ireland.

The United Kingdom is made up of England, Wales, Scotland and Northern Ireland.

The Republic of Ireland is a separate country with its own government. Until 1921 it was part of the UK.

B

C

D

0   200 km

Activities

1   Give a title to each of the maps **B**, **C** and **D**. Choose your titles from box **A** above.

2   Complete the crossword using the clues below. All of the answers can be found on map **E** opposite.

1   The Channel between England and France.
2   A Scottish river that starts in the Southern Uplands.
3   A capital city on the River Thames.
4   An island off the south coast of England.
5   A city on the River Clyde.
6   The capital city of Northern Ireland.
7   An island off the north coast of Wales.
8   A river with its source in the Cotswolds.
9   A city in the north east of Scotland.
10   The ocean west of Scotland.
11   A mountain range in Scotland.
12   The sea to the east of Britain.
13   A river flowing into the Bristol Channel.
14   A city in north east England.
15   A highland area of northern England.
16   The sea between England and Ireland.
17   The capital city of Wales.

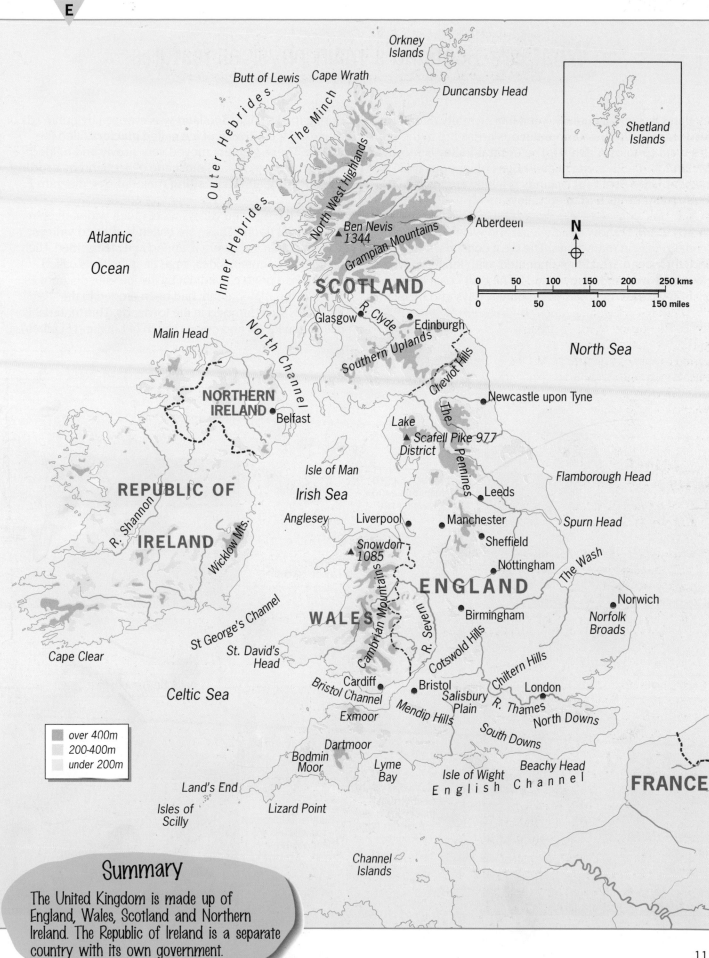

E

Orkney Islands

Butt of Lewis · Cape Wrath · Duncansby Head

Shetland Islands

Outer Hebrides

The Minch

North West Highlands

Atlantic Ocean

Inner Hebrides

Ben Nevis 1344

Aberdeen

Grampian Mountains

N

SCOTLAND

0 50 100 150 200 250 kms
0 50 100 150 miles

Glasgow R. Clyde Edinburgh

North Sea

Malin Head

North Channel

Southern Uplands

Cheviot Hills

NORTHERN IRELAND

Belfast

The Pennines

Newcastle upon Tyne

Lake District

Scafell Pike 977

REPUBLIC OF

Isle of Man

Irish Sea

Flamborough Head

R. Shannon

IRELAND

Anglesey Liverpool Manchester

Leeds

Spurn Head

Wicklow Mts.

Snowdon 1085

Sheffield

Nottingham

The Wash

Cambrian Mountains

ENGLAND

Norwich

St George's Channel

WALES

R. Severn

Birmingham

Norfolk Broads

Cape Clear

St. David's Head

Cotswold Hills

Chiltern Hills

Cardiff

Bristol

London

Celtic Sea

Bristol Channel

Salisbury Plain

R. Thames

North Downs

Exmoor

Mendip Hills

South Downs

Dartmoor

Bodmin Moor

Lyme Bay

Isle of Wight

Beachy Head

Land's End

English Channel

FRANCE

Isles of Scilly

Lizard Point

Channel Islands

over 400m
200-400m
under 200m

Summary

The United Kingdom is made up of England, Wales, Scotland and Northern Ireland. The Republic of Ireland is a separate country with its own government.

What are Scotland's main physical features?

Scotland is a small country, yet it has a variety of different **landscapes**. The most mountainous areas are in the north and west (photos **A** and **B**). The Central Valley is mostly lowland, but with some higher ridges of igneous rock. To the south, the Southern Uplands is a hilly area, with rounded summits and deep valleys.

Photos **C** and **D** show that the east and west coasts of Scotland vary in many ways. The west coast rises steeply from the ocean. It is deeply indented, with long narrow sea lochs or **fjords** and many islands. The east coast has few offshore islands, but many sand dunes, bays and headlands.

During the Ice Age, Scotland was covered by hundreds of metres of ice. 'Rivers' of ice called **glaciers** filled the valleys. These glaciers moved very slowly and helped shape the present day landscape of Scotland by **eroding**, **transporting** and **depositing** material. See diagram **E**.

After the Ice Age, the valleys were much wider, deeper and had flat floors. These are called **U-shaped valleys**. The mountain tops were rounded and steep sided. High up on the mountain-sides, large deep hollows called **corries** had been carved out by the ice (see diagram **F**). Much of the material that had been eroded in the highlands was dumped in the lowlands. This material is called **glacial till** and covers much of Lowland Scotland.

A Suilven

B Cairngorm, northern corries

C Loch Linnhe

D Pease Bay, south-east coast

North west – rugged mountains with narrow sea lochs and many islands

Cairngorms – high, rounded and steep

R. Spey

R. Dee

R. Tay

R. Forth

R. Clyde

R. Esk

Central Valley – mostly lowlands with higher ridges

Southern Uplands – rounded hills and deep valleys

N

0   100 km

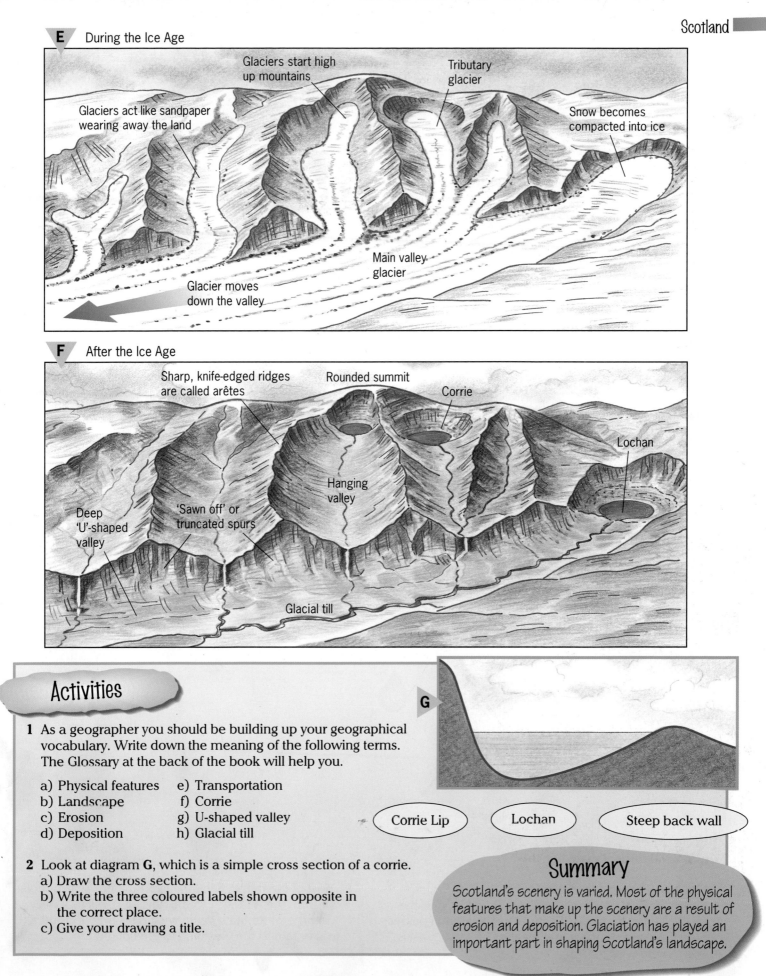

E During the Ice Age

Glaciers start high up mountains

Tributary glacier

Glaciers act like sandpaper wearing away the land

Snow becomes compacted into ice

Main valley glacier

Glacier moves down the valley

F After the Ice Age

Sharp, knife-edged ridges are called arêtes

Rounded summit

Corrie

Lochan

Hanging valley

Deep 'U'-shaped valley

'Sawn off' or truncated spurs

Glacial till

Activities

1 As a geographer you should be building up your geographical vocabulary. Write down the meaning of the following terms. The Glossary at the back of the book will help you.

a) Physical features   e) Transportation
b) Landscape   f) Corrie
c) Erosion   g) U-shaped valley
d) Deposition   h) Glacial till

2 Look at diagram **G**, which is a simple cross section of a corrie.
a) Draw the cross section.
b) Write the three coloured labels shown opposite in the correct place.
c) Give your drawing a title.

G

Corrie Lip   Lochan   Steep back wall

Summary

Scotland's scenery is varied. Most of the physical features that make up the scenery are a result of erosion and deposition. Glaciation has played an important part in shaping Scotland's landscape.

113

What is Scotland's climate like?

We talk about the weather so much in Scotland because our weather is very changeable. Records have been set for extremes of weather such as the wettest or coldest days and places. However, these extremes are unusual and it is rarely too hot or too cold or even too wet.

The word **climate** is used to describe the average weather of a place over many years. The climate of Scotland is described as **temperate** with warm summers, mild winters and some rain throughout the year. There are however, even in a small country like Scotland, differences in climate between one area and another.

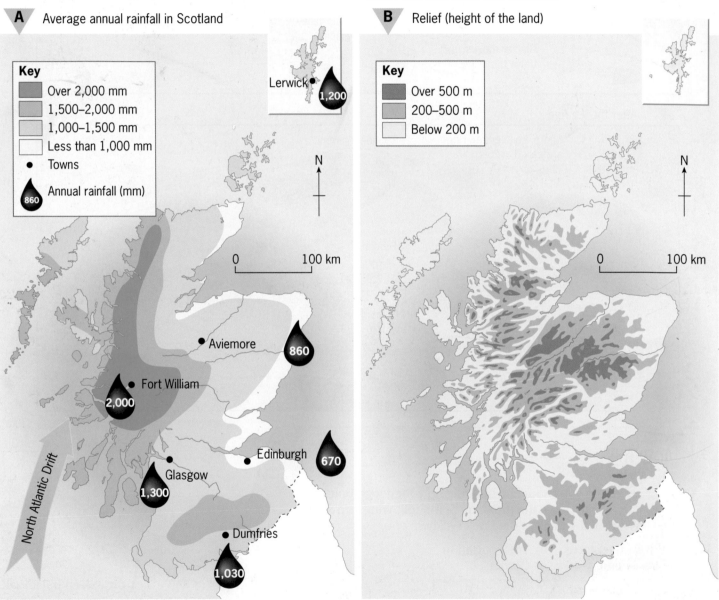

A Average annual rainfall in Scotland

Key
- Over 2,000 mm
- 1,500–2,000 mm
- 1,000–1,500 mm
- Less than 1,000 mm
- Towns
- Annual rainfall (mm) 860

Lerwick • 1,200

Aviemore • 860

Fort William • 2,000

North Atlantic Drift

Edinburgh 670

Glasgow 1,300

Dumfries 1,030

B Relief (height of the land)

Key
- Over 500 m
- 200–500 m
- Below 200 m

Why is Scotland's climate like this?

1 The North Atlantic Drift is an ocean current which brings warm water from the Caribbean Sea past the west coast of Scotland. This makes the air that passes over the current milder and means the west coast is less cold in winter.

2 Most of the rain in Scotland is brought by warm moist air from the Atlantic Ocean. This air cools as it rises over the mountains in the west. This brings cloud and heavy rain to the west and to the upland areas.

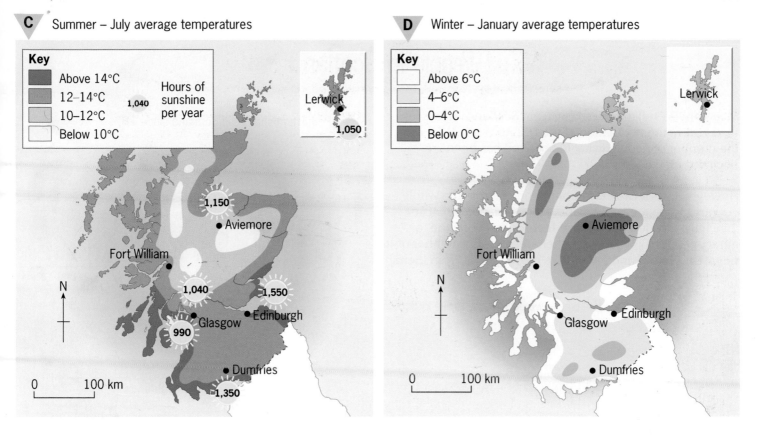

C Summer – July average temperatures

D Winter – January average temperatures

Activities

1 Make a larger copy of table **E**. Complete the table using the information from maps **A**, **C** and **D**. Write in the figures for each place.

2 Which of the named towns is:
   a) coldest in winter,
   b) warmest in summer,
   c) coolest in summer,
   d) wettest over a year,
   e) more likely to have sunshine?

E

Settlement	Annual rainfall (mm)	Summer temperature (°C)	Winter temperature (°C)	Sunshine (hrs)
Fort William				
Glasgow				
Lerwick				
Aviemore				
Edinburgh				
Dumfries				

3 Complete the statements by matching the heads on the left with their correct tails on the right.

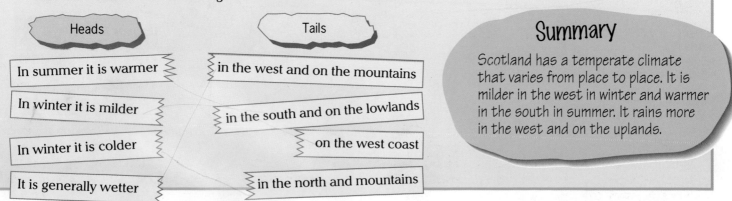

Heads

Tails

In summer it is warmer

In winter it is milder

In winter it is colder

It is generally wetter

in the west and on the mountains

in the south and on the lowlands

on the west coast

in the north and mountains

Summary

Scotland has a temperate climate that varies from place to place. It is milder in the west in winter and warmer in the south in summer. It rains more in the west and on the uplands.

People in Scotland

There are 5.1 million people living in Scotland. This is less than 10% of the population of the UK. As graph **A** shows, the number of people living in Scotland has been declining slowly since 1971.

However, this decline is not happening across the whole country. Some areas are gaining in population because they have a variety of **positive factors** that continue to attract people to live there. Other areas have a number of **negative factors** and people are moving away from them.

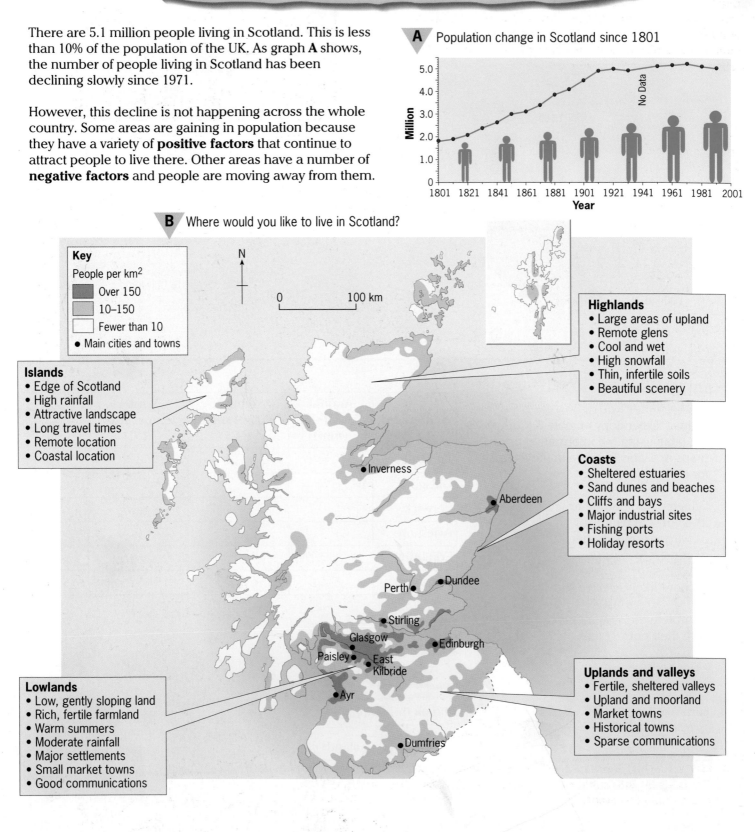

A Population change in Scotland since 1801

B Where would you like to live in Scotland?

Key

People per km²

- Over 150
- 10–150
- Fewer than 10
- ● Main cities and towns

Islands
- Edge of Scotland
- High rainfall
- Attractive landscape
- Long travel times
- Remote location
- Coastal location

Highlands
- Large areas of upland
- Remote glens
- Cool and wet
- High snowfall
- Thin, infertile soils
- Beautiful scenery

Coasts
- Sheltered estuaries
- Sand dunes and beaches
- Cliffs and bays
- Major industrial sites
- Fishing ports
- Holiday resorts

Lowlands
- Low, gently sloping land
- Rich, fertile farmland
- Warm summers
- Moderate rainfall
- Major settlements
- Small market towns
- Good communications

Uplands and valleys
- Fertile, sheltered valleys
- Upland and moorland
- Market towns
- Historical towns
- Sparse communications

People have been moving in and out of Scotland for thousands of years. The movement of people from one place to another is called **migration**. People who migrate into a country are called **immigrants** and those who migrate out are called **emigrants**. Migrations in the past have left their mark on place names (photo **C**), religions, language and architecture. More recent migration has given Scotland a **multicultural** society with a rich variety of beliefs, customs and traditions.

C Bilingual road sign on the Isle of Skye

D

My father left Scotland in 1945 seeking new opportunities in Canada. I was born in Vancouver but my close ties with Scotland led me to join the local pipe band.

My great grandfather left Ireland in 1846 to escape the potato famine. Five generations of our family have lived in Glasgow since.

I moved to Edinburgh in 1960 from Pakistan. I now run a successful business in Edinburgh and both my children are at University in Glasgow.

I live in Corby in England. My grandfather was one of three quarters of a million Scots who moved to England between 1840 and 1940.

My husband and I took early retirement from our jobs in London to work a small croft beside Loch Torridon.

Activities

1 Map **B** shows that the population of Scotland is not evenly distributed.
Copy chart **E** and write **five** more reasons for each factor.

E

Positive factors encouraging people into an area	Negative factors discouraging people from an area
fertile farmland	infertile soils

2 Make a list of where people in your class come from. Group them into suitable categories and plot the results on a bar graph.

3 Use the index of your atlas to find out which countries have settlements with the same names as these Scottish towns and cities.
Hamilton, Aberdeen, Glasgow, Perth

4 Read about the lives of the people shown in diagram **D**.
Choose **one** of the people and write a paragraph to expand their story.

Summary

Scotland's population has been decreasing slowly since 1971. People from many different countries live in Scotland but many Scots have emigrated to other parts of the world.

The best of Scotland

In 1998, 12 million tourists arrived in Scotland. They spent almost £2.5 billion which helped support many businesses and jobs in the tourist industry. Almost 2 million foreign visitors came from the USA, France and Germany.

Scotland has much to offer the visitor. It has a rich history and culture, along with interesting towns and cities. However, most foreign visitors come to Scotland for the scenery. Loch Lomond is a good example. The area of Loch Lomond is also a **honeypot**. Its popularity creates pressure on the landscape so it has been designated as Scotland's first National Park.

The holiday shown is typical of the sort of touring holiday many American visitors take to see Scotland.

A
Loch Lomond – Scotland's first National Park

Activities

1 List four features that make Scotland an interesting holiday destination.

2 Use chart **B** to measure the distances between the tour stops on the mainland. The first one has been done for you.

Ullapool to Inverness = 92 kilometres
Inverness to Aviemore =
Aviemore to Edinburgh =
Edinburgh to Galashiels =
Edinburgh to Glasgow =

3 Draw a star diagram like the one below to show at least six attractions of Edinburgh.

C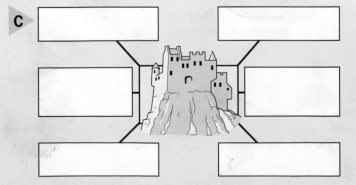

B

ABERDEEN							
153	AVIEMORE						
108	159	DUNDEE					
202	202	99	EDINBURGH				
260	260	157	53	GALASHIELS			
238	228	134	76	126	GLASGOW		
170	51	211	253	311	279	INVERNESS	
257	139	298	341	398	367	92	ULLAPOOL

Kilometres

4 Plan a seven day touring holiday for your family in Scotland.
- Prepare an itinerary as shown on page 119.
- Use the information on these two pages.
- A brochure from the Scottish Tourist Board would also be useful.
- If you have Internet access you could use the website at **http://www.calmac.co.uk** for ferry information.
- Try to add photographs, drawings and maps to your plan to make it more interesting.

D Caledonian MacBrayne carry more than five million passengers each year between the mainland and the Western Isles.

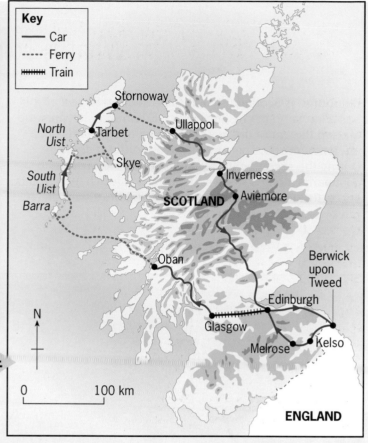

Key
— Car
- - - Ferry
++++ Train

Stornoway

North Uist Tarbet

Ullapool

Skye

Inverness

South Uist **SCOTLAND** Aviemore

Barra

Oban

Berwick upon Tweed

Edinburgh

Glasgow

Melrose Kelso

N

0   100 km

E

ENGLAND

Day 1   Arrive in Scotland
Arrive at Glasgow Airport from Boston at 9 am. Hire car and drive across Erskine Bridge, past Loch Lomond to Oban. Leave Oban on the 'Clansman' for the five hour sailing to Barra. Arrive in Barra 8 pm.

Day 2   Barra
Explore Barra. This is one of the most remote Western Isles. Catch the evening ferry for the 1½ hour sail to Loch Boisdale on South Uist.

Day 3   Drive to North Uist
Drive north to Lochmaddy using the new causeway linking Benbecula with North and South Uist. Until recently ferries had to be used.

Day 4   North Uist
Explore North Uist. This low-lying island has many tiny lochans and sea inlets. From Lochmaddy, ferries sail to Uig or Skye and to Leverburgh on Harris.

Day 5   Over the sea to Skye
A two island hop. Take the early morning ferry to Skye, one of the largest of the Scottish Islands. Famous for its excellent climbing in the Cuillin Hills. Visit Dunvegan Castle, the oldest inhabited castle in Britain. Catch the Hebrides for the two hour sailing to Tarbet, Harris.

Day 6   Stornoway
Drive 61 km to Stornoway and visit the ancient Standing Stones at Callanish. Leave Stornoway on the 7 pm sailing of the 'Isle of Lewis' to Ullapool. Drive to Inverness.

Day 7   Inverness
Morning free for shopping then drive to Aviemore near the Cairngorms. Aviemore is an all-year-round resort, catering for tourists in summer and skiers in winter. View the only reindeer herd in Britain.

Day 8   Edinburgh
Early morning drive on the A9/M90 and across the Forth Road Bridge to Edinburgh. Tour of 'Our Dynamic Earth' Exhibition, the new Scottish Parliament and in the afternoon, visit the former Royal Yacht Brittania. In the evening, view Edinburgh from high up on the ramparts of Edinburgh Castle.

Day 9   The Borders
Drive to Berwick-upon-Tweed. Follow the Tweed Valley west through the Abbey towns of Kelso and Melrose. Last night of holiday spent watching the Edinburgh Military Tattoo.

Day 10 Glasgow
Early morning train to Glasgow Queen Street Station. Taxi to Glasgow Airport in time for the 11 am flight to Boston.

Summary
Scotland's scenery, towns and history attract many visitors. Transport improvements have made remote areas more accessible in recent years.

What is Edinburgh like?

Edinburgh is Scotland's capital and second largest city. It has a long and interesting history. Edinburgh first grew up around the castle that was built upon the remains of an ancient volcano. This provided a good **defensive site**. During the seventeenth century, the city slowly spread eastwards along a steep and narrow ridge. There was little space for development here so the buildings were tall and packed closely together.

In the eighteenth century a new town was built on flat land to the north where there was space for wide streets and fine buildings. Since then, the city has grown gradually outwards in all directions. New housing estates have been built, industrial areas developed, and large shopping complexes constructed on the outskirts.

Many older areas are being redeveloped. One such area is the Leith Docklands shown in the photo below and map opposite.

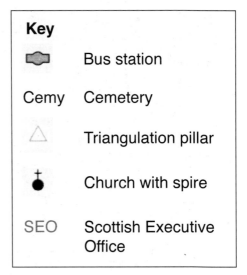

Key

⬭	Bus station
Cemy	Cemetery
△	Triangulation pillar
♀	Church with spire
SEO	Scottish Executive Office

Activities

1 Draw and name the symbols shown at these grid references. Answer like this:

254772 = △ = lighthouse:

| 258742 | 262756 | 272761 | 263743 |

2 Name the features at these locations.

| 252735 | 269739 | 260739 | 245755 |

3 The photo above was taken from Calton Hill (263743) looking north. Match the four locations **A**, **B**, **C** and **D** on the photo with the names below found on the map.

| Leith Docks | Hillside | Leith Links | Mills |

4 The map includes Edinburgh's city centre. Where do you think it is?
   a) Write a four figure grid reference to locate the centre.
   b) Give two reasons for your answer.

5 The new Scottish Executive Office is located on the site of an old dock at 267768.
   Suggest one advantage and one disadvantage of this site.

Summary

Edinburgh is a large city that is crowded and always busy. Redevelopment schemes are helping improve conditions in the older areas.

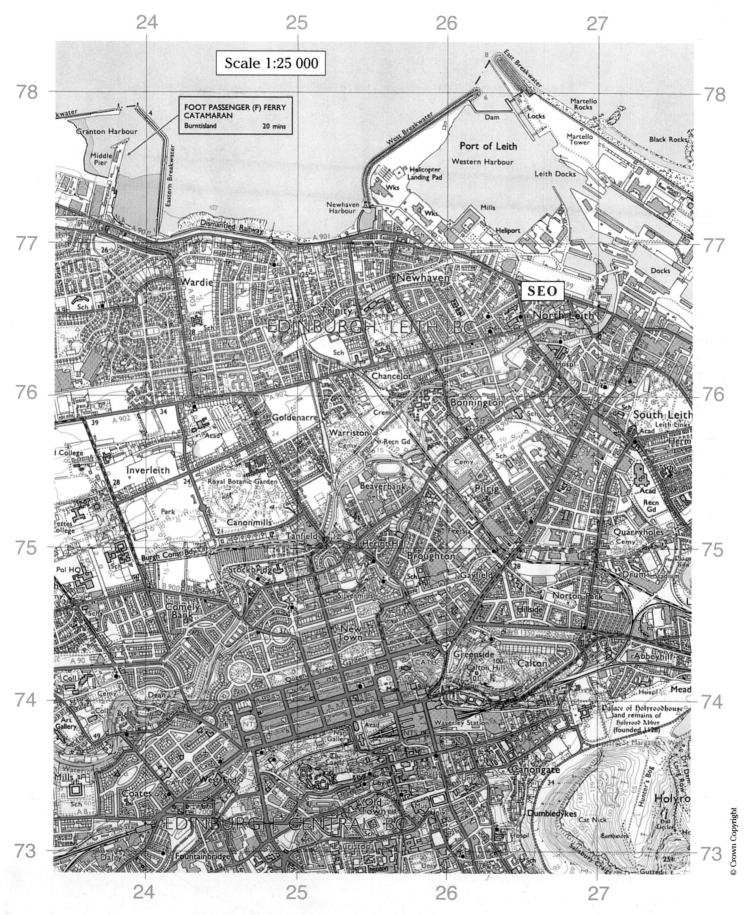

Scale 1:25 000

FOOT PASSENGER (F) FERRY
CATAMARAN
Burntisland   20 mins

The growth of Edinburgh

Edinburgh is located on an area of mainly lowland between the Firth of Forth to the north and Pentland hills to the south. The city has grown rapidly in recent years and now stretches some 20 km from one side to the other. Many smaller towns and villages like Leith,

Cramond and Colinton are now part of Edinburgh's built up area. There are two main reasons for Edinburgh's growth. The first is its good location and the second its many **functions**. Some of these are shown below.

Edinburgh Business Park

Edinburgh Castle

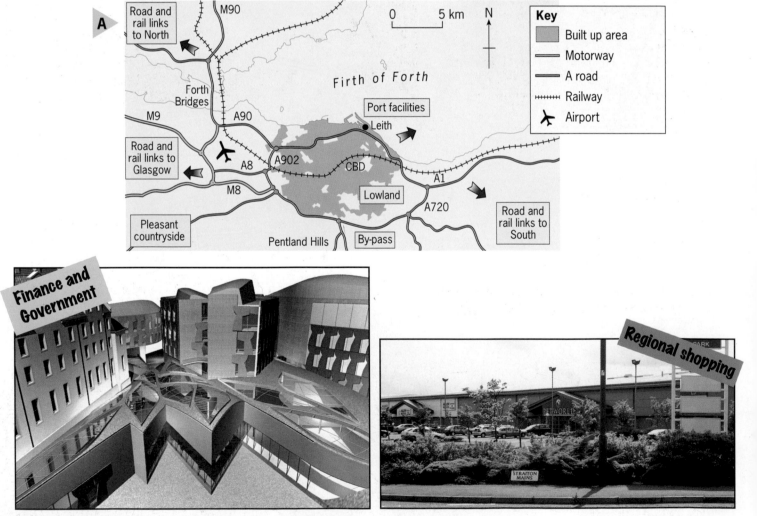

An artist's impression of the completed Scottish Parliament buildings

Straiton shopping centre

Like all big cities, Edinburgh has its problems. Many people think that most of these problems are due to growth that has been too rapid.

One of the problems is a lack of space leading to high land prices. As graph **B** shows, Edinburgh is now one of the most expensive office locations in the world. High prices like these can encourage businesses to locate elsewhere and reduce economic growth. Diagram **C** shows two other problems and possible solutions.

B

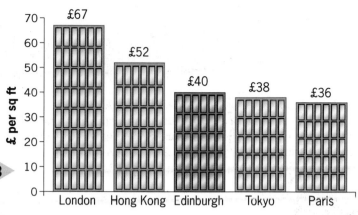

£ per sq ft

London £67
Hong Kong £52
Edinburgh £40
Tokyo £38
Paris £36

C

EDINBURGH'S PROBLEMS

Lack of space

Problem
Most of the useable land in the city centre and suburbs has already been built on. The green belt around the city has restricted out-of-town developments.

Solution
Old or derelict sites in the city centre have been cleared for new building. Land use changes have been made to allow developments such as the Parliament buildings and Financial district to be built. Relaxation of green site regulations have allowed planned developments on the city edge.

Traffic congestion

Problem
With no rail based rapid transit system, most of Edinburgh's commuters have to travel by car or bus. Over 100,000 vehicles enter the city each day. At times this causes **traffic grid lock**.

Solution
The council have encouraged the use of public transport by improving bus travel, introducing 'park and ride' schemes and reducing car parking spaces in the city centre. A by-pass helps to keep through traffic out of the city.

Activities

1 Give six location factors from map **A** that have helped Edinburgh's growth. Divide your list into 'transport' and 'others'.

2 A function is something that a city provides. Complete the star diagram below (**D**).

D

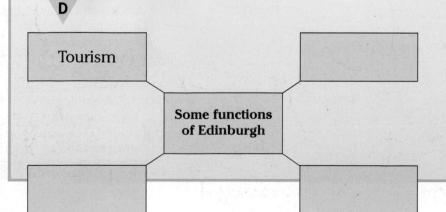

Tourism

Some functions of Edinburgh

3 Draw a poster to discourage motorists from taking their cars into Edinburgh city centre.
● Include reasons why they should not use their cars.
● Show advantages of using public transport.
● Colour your poster and make it interesting and attractive.

Summary

Edinburgh has grown rapidly in recent years because of its good location and many important functions. The growth of the city has caused problems for the people living there.

What are the Western Isles like?

A Castlebay, Barra's main settlement

The Western Isles, or Eilean Siar in gaelic, is one of the 32 local authority areas into which Scotland is divided. Each authority organises its own local services such as education, housing and health.

Despite having one of Scotland's most distinctive and attractive landscapes, the Western Isles is an area of low **population density** in which it can be difficult to make a living.

Barra (Barraigh) is a small, **remote** island with a population of 1,250 at the southern end of the Western Isles. This rugged, glaciated island rising nearly 400 metres out of the Atlantic Ocean is a beautiful place with its sandy bays in the west and rocky cliff coastline in the east. Despite the difficulties shown on diagram **B**, the population is no longer declining. Although the majority of jobs on the mainland are found in the **service** sector, fishing, tourism and **crofting** all provide employment on this most westerly island.

The small island of Vatersay (Bhatarsaigh) was joined to Barra in 1992 by a causeway (photo **D**).

WESTERN ISLES TOURIST BOARD

BORD TURASACHD NAN EILEAN

B Barra's environment

Steep, rocky hills dominate the island

Thin, stony, infertile, acid soils on the hills

Warm, wet summers and mild, wet winters

Gale force winds from the Atlantic for 35 days a year

Five hour vehicle ferry link to Oban

Two flights a day to Glasgow from Traigh Mhor beach

Sandy bays and dunes on the west coast

High, rocky cliffs on the east coast

Narrow roads with passing places

Small areas of fertile sandy **machair**

Small crofting townships with shared **common grazing land**

Hello,

I'm Mhairi and I live with my brother and parents on one of the 40 **crofts** in Eoligarry (Eolaigearraidh) in the north of Barra. All crofts on the island are part time because the infertile soils, steep slopes and wet, windy weather make farming difficult.

The hilly **common grazing land** is shared by all the crofters in the township. Thirteen of the crofts are worked by pensioners and others belong to teachers, fishermen, a garage owner and a builder. Only three are used as holiday homes.

We rent four hectares of 'machair' — the low, flat, sandy and more fertile land by the coast. We grow potatoes for our own use and hay for winter feed. We keep hens and three cows for eggs, milk and meat.

Land is rented on a nearby uninhabited island (Gighay) where we keep 150 sheep. In the summer we go across to dip and shear them — great fun but hard work. We sell our sheep and cattle in Oban in November.

My dad owns a fishing boat which supplies the fish processing plant at Ardveenish with prawns and scallops. The factory employs 40 people and has five lorries a week transporting shellfish to the main markets in France and Spain.

I have to travel 15 km each day to Castlebay High School which has 102 pupils, but my brother attends the 22 pupil Eoligarry Primary School. We enjoy using the sports centre and swimming pool in Castlebay and look forward to special shopping trips to Oban — if the sea is calm.

C Population of Vatersay

1911	𝕏𝕏𝕏𝕏𝕏𝕏𝕏𝕏𝕏𝕏𝕏𝕏	288
1989	𝕏𝕏𝕏	59
1995	𝕏𝕏𝕏𝕏	83

🧍 = 25 people

D Vatersay Causeway

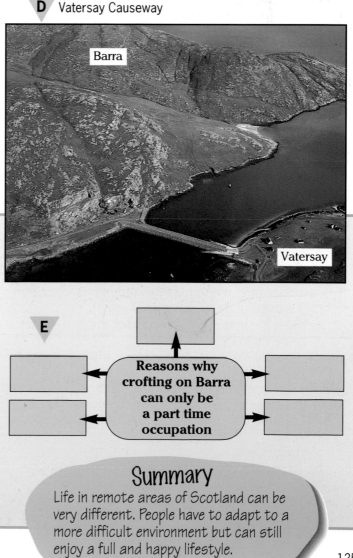

Barra

Vatersay

Activities

1 All 400 crofts on Barra are run as part time farms.
   a) Write down **five** other jobs that crofters may do to help them earn a living.
   b) Write down **five** things that make crofting different from other types of farming you have studied in Scotland.

2 Look at graph **C** and photo **D**.
   Vatersay is a small island 250 metres south of Barra.
   a) Write down **three** reasons why so many people left Vatersay.
   b) Explain why the population has risen since 1989.

3 Copy and complete diagram **E**.
   Try to write less than eight words in each box.

E

Reasons why crofting on Barra can only be a part time occupation

Summary

Life in remote areas of Scotland can be very different. People have to adapt to a more difficult environment but can still enjoy a full and happy lifestyle.

125

Website http://**www.Scotland-info.co.uk/barra.htm** gives more information on Barra

Would you like to visit the Western Isles?

One of the reasons the population on Barra has declined less in recent years has been an increase in tourism. In 1999 over 14,000 passengers used the ferry from Oban and 80% of these were in the summer months. Over 8,000 people visit the Tourist Information Centre in Castlebay each year.

Tourists also make good use of the 14-seater plane that flies in twice a day from Glasgow. The plane lands on the beach at Traigh Mhor; the most westerly airfield in the UK. The plane can only land at low tide and therefore flight times change daily. The weather, especially strong winds, disrupts about 80 flights every year.

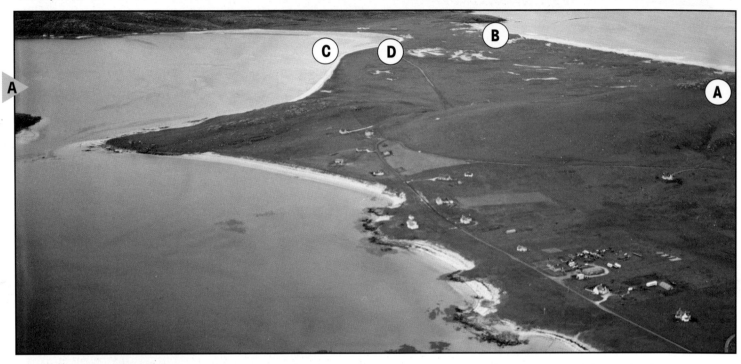

Activities

1 Look at the OS map. Name the crofting townships at:
   a) 7099   b) 6601   c) 7007.

2 Castlebay (grid reference 6698 and photo **A** on page 124) is the main settlement on Barra.
   a) Using the map write down **three** things a visitor could do in the town.
   b) Using the map give **two** reasons why this is a good site for a settlement.

3 Photo **A** was taken from above grid square 7107 looking towards the south west. Use the OS map to name:
   a) The hill at **A**.
   b) The sand feature at **B**.
   c) The beach at **C**.
   d) The height of the land at **D**.

4 Find the highest point on Barra.
   a) Name the mountain.
   b) Write down its height.
   c) Give its grid reference.

5

Donald and Eileen have just left school and have different opinions about their future on Barra.

Work in pairs, taking the roles of Donald and Eileen. Write a postcard to each other giving good reasons to persuade the other to change their mind.

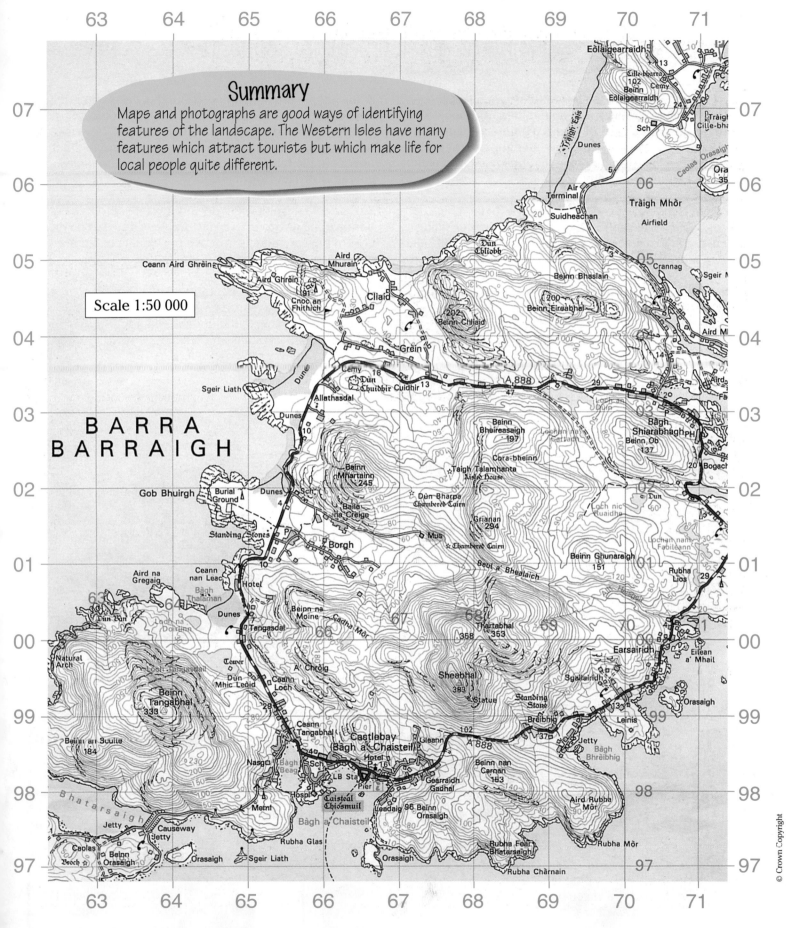

Summary
Maps and photographs are good ways of identifying features of the landscape. The Western Isles have many features which attract tourists but which make life for local people quite different.

Scale 1:50 000

BARRA
BARRAIGH

Sustainable development in Scotland

The UK is one of the world's wealthiest countries. Industrial development has resulted in many people having a good **standard of living**. This increase in wealth combined with more leisure time and improvements in transport has led to an increase in tourism and recreation in both summer and winter.

Scotland is the only location in the UK with land high enough to receive sufficient snowfall suitable for skiing. There are five main centres, attracting half a million skiers a year, providing 3,000 jobs and an income of £30 million.

One of the main centres is in the Cairngorm Mountains. This centre first opened in 1961 and has since extended to provide 50 km of ski runs for 200,000 skiers a year.

Many people are concerned about this development. They argue that the aim of development should be to improve conditions for people. In doing this, however, development should not harm the environment either now or in the future. Development like this is called **sustainable development**.

Sustainable development does not waste resources or damage the environment. It is progress that can go on year after year. It helps improve our quality of life but does not spoil our chances in the future.

Conservationists are concerned that the popularity of the Cairngorms in both summer and winter will result in irreparable damage to the unique and fragile sub-arctic landscape and its plants and wildlife.

A Attractions of the Cairngorms and the resulting problems

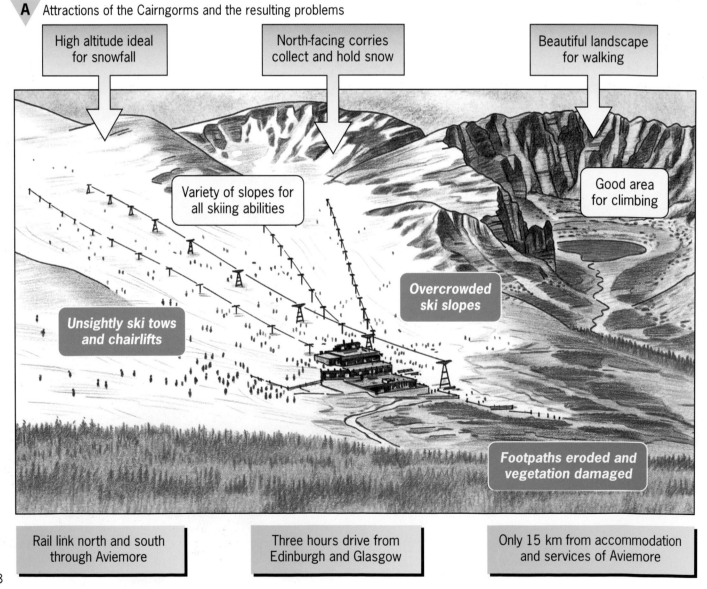

High altitude ideal for snowfall

North-facing corries collect and hold snow

Beautiful landscape for walking

Good area for climbing

Variety of slopes for all skiing abilities

Overcrowded ski slopes

Unsightly ski tows and chairlifts

Footpaths eroded and vegetation damaged

Rail link north and south through Aviemore

Three hours drive from Edinburgh and Glasgow

Only 15 km from accommodation and services of Aviemore

B Will the planned funicular railway make development more sustainable?

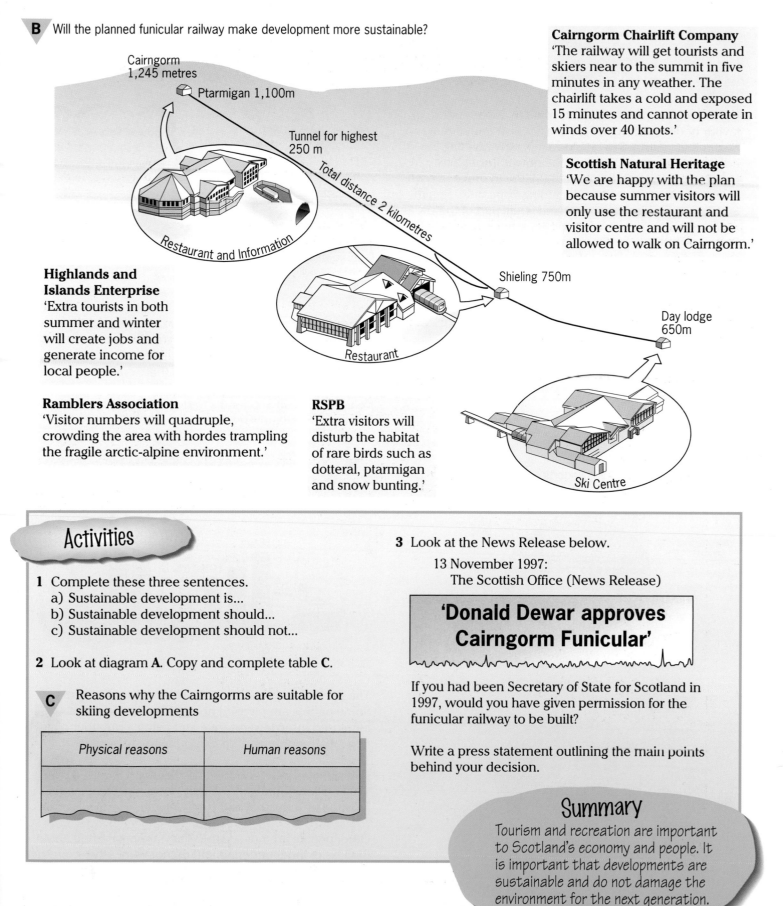

Cairngorm
1,245 metres

Ptarmigan 1,100m

Tunnel for highest
250 m

Total distance 2 kilometres

Restaurant and Information

Shieling 750m

Day lodge
650m

Restaurant

Ski Centre

Cairngorm Chairlift Company
'The railway will get tourists and skiers near to the summit in five minutes in any weather. The chairlift takes a cold and exposed 15 minutes and cannot operate in winds over 40 knots.'

Scottish Natural Heritage
'We are happy with the plan because summer visitors will only use the restaurant and visitor centre and will not be allowed to walk on Cairngorm.'

Highlands and Islands Enterprise
'Extra tourists in both summer and winter will create jobs and generate income for local people.'

Ramblers Association
'Visitor numbers will quadruple, crowding the area with hordes trampling the fragile arctic-alpine environment.'

RSPB
'Extra visitors will disturb the habitat of rare birds such as dotteral, ptarmigan and snow bunting.'

Activities

1 Complete these three sentences.
a) Sustainable development is...
b) Sustainable development should...
c) Sustainable development should not...

2 Look at diagram **A**. Copy and complete table **C**.

C Reasons why the Cairngorms are suitable for skiing developments

Physical reasons	Human reasons

3 Look at the News Release below.

13 November 1997:
The Scottish Office (News Release)

'Donald Dewar approves Cairngorm Funicular'

If you had been Secretary of State for Scotland in 1997, would you have given permission for the funicular railway to be built?

Write a press statement outlining the main points behind your decision.

Summary
Tourism and recreation are important to Scotland's economy and people. It is important that developments are sustainable and do not damage the environment for the next generation.

How interdependent is Scotland?

In 1998 the people of Scotland voted in a referendum in favour of **devolution**. In 1999 the Scottish Parliament became established to be based in a new building in Edinburgh. Scotland now has the power through its Members of Scottish Parliament (MSPs) to make its own decisions on many matters including Education, Health and Transport. However, Scotland is still part of the UK and represented by MPs at Westminster where decisions on Defence and Foreign Affairs are made.

Since 1973 the UK, and therefore Scotland, has been a member of the European Union (EU). In 2000 there were 15 member countries but with several other countries wishing to join there could be another 13 members by 2010.

A population of over 370 million makes the EU one of the largest trading groups in the world. This has provided Scotland with a large and accessible market for her **imports** and **exports**.

Did you know?
In Scotland £12.14 per person per day is collected for Government expenditure. Of which 37 pence goes to the EU.

All countries need help from other countries if they are to make progress and improve their standards of living. Countries that work together are said to be **interdependent**. One of the ways that Scotland is interdependent with the UK and Europe is through **trade**. Nearly 60% of all Scotland's trade is with the rest of the UK and today Scotland's trade is worth £20 billion a year.

A Scotland's exports and imports

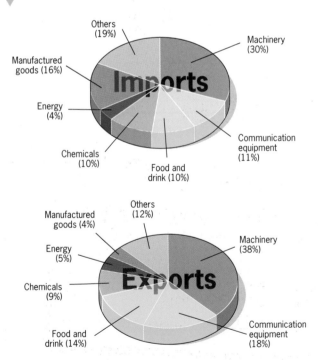

Imports

- Others (19%)
- Manufactured goods (16%)
- Energy (4%)
- Chemicals (10%)
- Food and drink (10%)
- Communication equipment (11%)
- Machinery (30%)

Exports

- Manufactured goods (4%)
- Others (12%)
- Energy (5%)
- Chemicals (9%)
- Food and drink (14%)
- Communication equipment (18%)
- Machinery (38%)

B Scotland's export trade (excluding the UK)

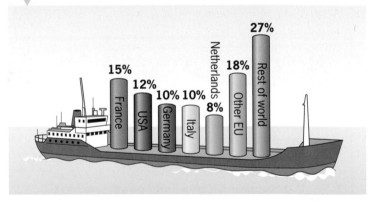

- France 15%
- USA 12%
- Germany 10%
- Italy 10%
- Netherlands 8%
- Other EU 18%
- Rest of world 27%

The EU has developed into more than a trading group. Some people are pleased about this. They feel that increased **interdependence** and being a member of a rich and important group can bring many benefits.

For instance, between 1994–99, Scotland received over £700 million pounds of EU aid to help redevelop areas affected by industrial and rural decline.

Other people are not so sure. They are concerned about the loss of independence and do not want other countries to interfere in their laws and way of life. Others worry that Scotland puts more money into the EU than it gets back. These problems may get worse if some of the poorer countries of Eastern Europe join the EU.

Ways in which the EU is helping Scotland

C

Stornoway Harbour
The new RoRo Berth and Passenger Terminal at Stornoway Harbour cost £8 million. EU money helped to finance this improvement in transport.

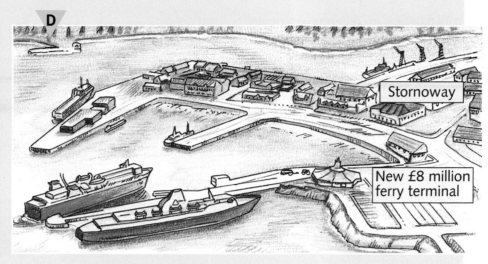

D

Stornoway

New £8 million ferry terminal

Scottish Borders
£217,000 from EU RETEX Fund. Forty textile firms employing 3,000 people are suffering from foreign competition. This money is helping Scottish Borders Enterprise and the textile firms to promote the international reputation of cashmere as a fashion product. This should increase demand.

University of the Highlands and Islands
EU is contributing £10 million to this £35 million project. With 10 main campuses spread across the Highlands and a network of resource centres it should help stem the migration of young talent from the region.

E

Activities

1 a) What is meant when a country is said to be interdependent?
b) Why does a country have to trade?

2 Look at diagram **F**.
a) Construct a bar graph to show these figures.
b) Write down one advantage and one disadvantage for Scotland having this large number of overseas companies.

3 Do you think membership of the EU is good or bad for Scotland? Give reasons and actual examples to support your answer.

F

Number of overseas companies located in Scotland			
American	280	Norwegian	28
French	78	Swedish	25
Japanese	62	Swiss	23
German	50	Danish	16
Canadian	30	Rest of world	77
Dutch	28		

Summary
Being a member of the European Union has helped Scotland to develop trade and become interdependent. Scotland now has more independence within the UK.

How developed is Scotland?

All countries are different. Some are rich and have high standards of living. Others are poor and have lower standards of living. Countries that differ in this way are said to be at different levels of **development**.

There are many ways of measuring development. Map **A** compares standards of living in the EU by wealth. A fuller picture of a country is given by looking at employment, health, education, trade and population statistics. Scotland has very similar statistics to the UK and these are shown on pages 8 and 11.

Most people agree that Scotland is a modern, wealthy and highly developed nation. However, as in most other countries, wealth and a high standard of living are not shared equally between everyone. Many people are well off and live an enjoyable life but some are poor and live in difficult conditions. For instance, 4,000 people in Scotland live in hostels and poor quality accommodation called squats.

A Comparison of wealth in the EU

0   1000 km

N

Key

GDP per person where EU average is 100

Over 125   More developed

101–125

75–100

Less than 75   Less developed

Activities

1   Look at table **B**.
Do you think Scotland's standard of living is **poorer/similar to/better than** the rest of the UK? Give reasons for your answer.

2   Look at diagram **C**.
Suggest two other industrial jobs the older man may have worked in and two that his son could work in today in Scotland.

3   Imagine that you are employed by Lanarkshire Development Agency. A computer monitor company from the USA is looking for a site for a new factory. Design a brochure or poster which could be sent to them which outlines the benefits of locating at Eurocentral.
Use the internet to gather information. Start with **http://www.lda.co.uk**

B

Indicator	Scotland	Rest of UK
Births (per 1,000 people)	11.6	12.3
Deaths (per 1,000 people)	11.6	11.0
Infant deaths (per 1,000 people)	5.3	6.1
Cars (per 1,000 people)	209	219
Literacy rate (%)	99	99
Unemployment rate (%)	5.2	4.9

C

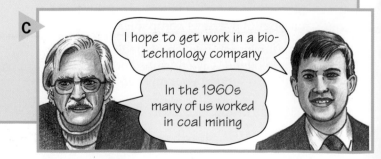

I hope to get work in a bio-technology company

In the 1960s many of us worked in coal mining

Scotland developed throughout the nineteenth and early twentieth centuries as a major industrial nation. From the 1960s many old traditional industries declined and closed. This resulted in some areas of Scotland suffering from high unemployment, low income, a lower standard of living and therefore a poorer quality of life.

During the last 20 years great efforts have been made to attract and encourage new industries to regions like Lanarkshire, where these problems are greatest. Diagram **D** shows one of these new development areas at Eurocentral. New jobs have reduced unemployment, raised standards of living and given people in deprived areas new hope for the future.

D   **Eurocentral development**

euro**central**

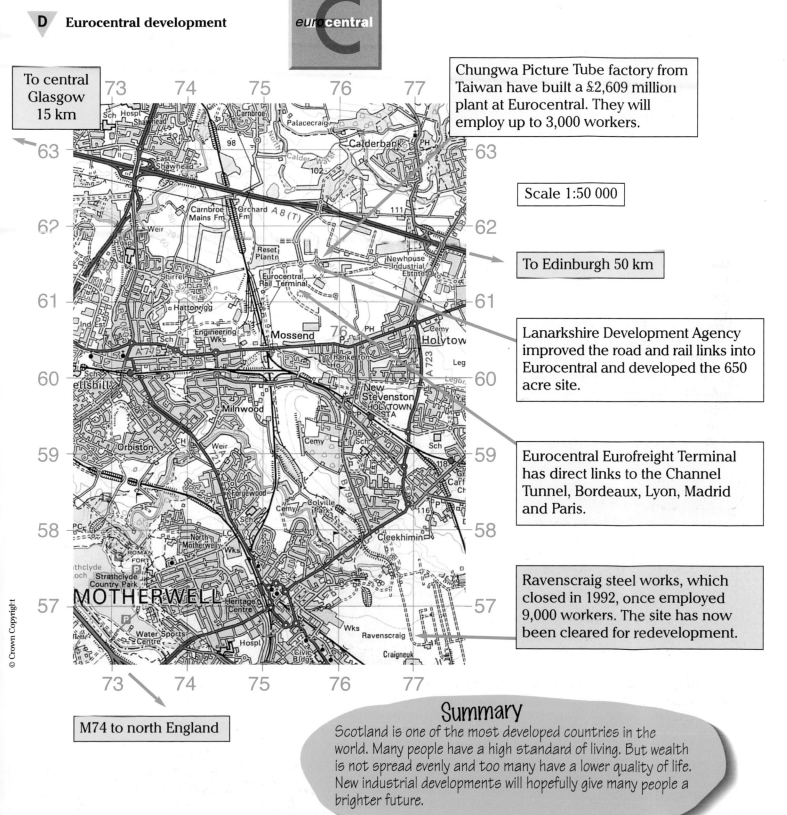

To central Glasgow 15 km

Chungwa Picture Tube factory from Taiwan have built a £2,609 million plant at Eurocentral. They will employ up to 3,000 workers.

Scale 1:50 000

To Edinburgh 50 km

Lanarkshire Development Agency improved the road and rail links into Eurocentral and developed the 650 acre site.

Eurocentral Eurofreight Terminal has direct links to the Channel Tunnel, Bordeaux, Lyon, Madrid and Paris.

Ravenscraig steel works, which closed in 1992, once employed 9,000 workers. The site has now been cleared for redevelopment.

M74 to north England

Summary

Scotland is one of the most developed countries in the world. Many people have a high standard of living. But wealth is not spread evenly and too many have a lower quality of life. New industrial developments will hopefully give many people a brighter future.

Glossary

A

Active volcano	A volcano that has erupted recently and is likely to erupt again. *66, 88*
Ageing population	An increasing number of older people in a country's population structure. *91*
Aid	Help usually given by the richer countries, international agencies and voluntary organisations mainly to poorer countries. *12, 34*
Amerindian	A member of an Indian tribe of North, Central or South America. *14*
Amazon	Brazil's greatest river and the second longest in the world. *16*

B

Beach resort	A hotel next to the sea which provides everything a tourist needs, e.g. restaurants, water sports, entertainments. *45, 47*
Birth rate	The number of people born for each 1000 of the population. *8, 43, 52, 91*

C

Cash crop	A crop that is grown for profit and usually exported. *54*
Carnival	A street festival with parades, parties, music, dancing and fancy dress. *20*
Climate	The average weather conditions of a place. *114*
Commercial farming	When farm produce is sold for a profit. *22, 54*
Commercial logging	The chopping down of trees which are then sold for a profit. *36*
Common grazing land	Land belonging to a crofting community which is shared by all crofters and used for grazing sheep and cattle. *125*
Communications	The ways in which people, goods and ideas move from one place to another. *78*
Commuter	A person who lives in one place and travels some distance to work in another. *97*
Conservation	The protection of the environment. *36*
Coral	A type of limestone rock made up of the skeletons of tiny marine creatures. *40, 41, 45, 47*
Coral reef	A band of coral lying off the coast which forms a fragile environment and protects the coast. *40, 41, 45, 47*
Corrie	Steep-sided hollow left after a glacier melts. *112, 113*
Crofting	A type of small scale subsistence farming found in the highlands and Islands of Scotland. *124, 125*

D

Death rate	The number of people who die per 1000 of the population. *8, 9, 43, 91*
Defensive site	A place where a settlement grew up because it was easy to defend in olden times. Often on a hilltop, or in the neck of a meander. *120*
Deposition	The laying down of material carried by rivers, sea, ice or wind. *112, 113*
Densely populated	An area that is crowded with people. *19, 64, 78, 120, 121*
Development	Development means growth. It involves changes and usually brings about improvement. Countries can be at different stages of development depending on how 'rich' or how 'poor' they are. *4, 6, 7, 11, 13, 32, 33, 60, 84, 85, 108, 109, 132, 133*
Developed country	A country which has a lot of money, many services and a higher standard of living. *7, 10, 32, 60, 108, 109, 132, 133*
Developing country	A country which is often quite poor, has few services and a lower standard of living. *7, 10, 32, 33, 48, 50, 58, 59, 60, 108, 109, 132, 133*
Development indicators	Methods used to measure development. *8, 9, 10, 11, 33, 84, 108, 109*
Devolution	Setting up of national governments within the UK system, e.g. Wales, Scotland. *130*
Dormant volcano	A volcano that has erupted in the last 2000 years but not recently. *88*

E

Earthquake	A movement, or tremor, of the earth's crust. *66, 88, 89, 96*
Earth's crust	The thin outer layer of the earth. *66*
Economic indicators	Measures of development that are based on wealth, e.g. GNP, trade and energy use. *7, 33*
Economic Union (EU)	A group of European countries working together for the benefit of everyone. *82, 83*
Economic miracle	The term given to Brazil's rapid industrial growth and great increase in wealth. It occurred in the 1960s and 1970s. *36*
Emigrant	A person who moves away from a country. *117*
Erosion	The wearing away and removal of rock, soil, etc. by rivers, sea, ice and wind. *112, 113*
Ethnic groups	People with similar culture, background and way of life. *42*
Exports	Goods sold to other countries. *34, 58, 59, 82, 106, 107, 130, 131*
Extinct volcano	A volcano that has not erupted in historic times and is not expected to erupt again. *88*